Improving Concrete and Mortar Using Modified Ash and Slag Cements

Improving Concrete and Mortar Using Modified Ash and Slag Cements

Leonid Dvorkin, Vadim Zhitkovsky
National University of Water and Environmental Engineering
Rivne, Ukraine

Mohammed Sonebi
Queen's University of Belfast
Belfast, United Kingdom

Vitaliy Marchuk, Yurii Stepasiuk
National University of Water and Environmental Engineering
Rivne, Ukraine

CRC Press is an imprint of the
Taylor & Francis Group, an **informa** business

CRC Press
Taylor & Francis Group
6000 Broken Sound Parkway NW, Suite 300
Boca Raton, FL 33487-2742

© 2020 by Taylor & Francis Group, LLC
CRC Press is an imprint of Taylor & Francis Group, an Informa business

No claim to original U.S. Government works

International Standard Book Number-13: 978-0-367-46348-9 (Hardback)

This book contains information obtained from authentic and highly regarded sources. Reasonable efforts have been made to publish reliable data and information, but the author and publisher cannot assume responsibility for the validity of all materials or the consequences of their use. The authors and publishers have attempted to trace the copyright holders of all material reproduced in this publication and apologize to copyright holders if permission to publish in this form has not been obtained. If any copyright material has not been acknowledged, please write and let us know so we may rectify in any future reprint.

Except as permitted under U.S. Copyright Law, no part of this book may be reprinted, reproduced, transmitted, or utilized in any form by any electronic, mechanical, or other means, now known or hereafter invented, including photocopying, microfilming, and recording, or in any information storage or retrieval system, without written permission from the publishers.

For permission to photocopy or use material electronically from this work, please access www.copyright.com (http://www.copyright.com/) or contact the Copyright Clearance Center, Inc. (CCC), 222 Rosewood Drive, Danvers, MA 01923, 978-750-8400. CCC is a not-for-profit organization that provides licenses and registration for a variety of users. For organizations that have been granted a photocopy license by the CCC, a separate system of payment has been arranged.

Trademark Notice: Product or corporate names may be trademarks or registered trademarks, and are used only for identification and explanation without intent to infringe.

Visit the Taylor & Francis Web site at
http://www.taylorandfrancis.com

and the CRC Press Web site at
http://www.crcpress.com

Contents

Preface vii
Authors ix

Chapter 1   Composite Ash-Containing Cements and Effective Concrete
Based on Them 1

1.1   Fly Ash—An Active Filler of Cement Systems 1
1.2   Pozzolanic Activity, Hydration Feature, and Rheological
Properties of Cement-Ash Pastes with Additives Modifiers 7
    1.2.1   Pozzolanic Activity 7
    1.2.2   Features of Hydration 9
    1.2.3   Rheological Properties 11
1.3   Structure Formation and Phase Composition of
Ash-Containing Composite Concrete 14
    1.3.1   The Initial Structure Formation 14
    1.3.2   Phase Composition and Porosity of Cement Stone 18
1.4   Construction and Technical Properties of Modified Ash
Composite Cements and Mortars Based on Them 22
    1.4.1   Material Composition 22
    1.4.2   Properties of Mortar Mixtures 24
    1.4.3   Strength Indices of Ash-Containing Composite
Cements 33
    1.4.4   The Adhesive Properties of Mortars Based on Ash
Containing CC 37
1.5   High-Tech Concretes Based on Ash-Containing
Composite Cement 40
    1.5.1   Properties of Concrete Mixtures 40
    1.5.2   The Strength of Concrete 44
    1.5.3   Deforming Properties 52
    1.5.4   Crack Resistance 54
    1.5.5   Water Impermeability 55
    1.5.6   Frost Resistance 56
    1.5.7   Designing Compositions of High-Tech Concrete 57
    1.5.8   Example of Calculating the Composition
of Concrete Type HPC 58

| **Chapter 2** | Activated Low Clinker Slag Portland Cement and Concrete on Its Basis .... 61 |

2.1 Activation of Slag Binders 61
    2.1.1 Methods for Activating Slag Binders 61
    2.1.2 Activation of the Low Clinker Slag Portland Cement (LSC) 64
    2.1.3 Integrated Methods of Activation of LSC 72
    2.1.4 Features of Hydration and Structure Formation of the LSC 77
2.2 Normal-Weight Concretes Based on Activated LSC 82
    2.2.1 Calculation Example 89
    2.2.2 Foam Concrete 92
    2.2.3 Vibropressed Sawdust Concrete 96
2.3 Fiber-Reinforced Concrete Based on Activated LSC 98
    2.3.1 Fiber Concrete at LSC 98
    2.3.2 Calculation Example 103
2.4 Frost Resistance of Concrete and Fiber-Reinforced Concrete and Corrosion Resistance in Concrete at LSC 106
    2.4.1 Frost Resistance of Concrete and Fiber-Reinforced Concrete Made Based on LSC 106
    2.4.2 Corrosion Resistance in Concrete at LSC 108
2.5 Dry Building Mixes and Mortars Based on Activated LSC 114
    2.5.1 Dry Mixes for Masonry Mortars 114
    2.5.2 Calculation Example 118
    2.5.3 Dry Mixes for Porous Masonry Mortars 121
    2.5.4 Dry Mixes for Self-leveling Floors 122

Chapter 3 Dry Construction Mixtures and Mortars Based on Them Using the Dust of Clinker-Burning Furnaces 127

3.1 Clinker Kiln Dust Is an Active Component of Cementitious Systems 127
3.2 Rheological and Structural-Mechanical Properties of Water Pastes of Dust-Slag-Superplasticizer and Dust-Slag-Cement-Superplasticizer Systems 131
3.3 Technological Parameters of Obtaining Composite Cement-Dust-Slag Binders of Low Water Demand 146
3.4 Dry Mixes and Mortars Based on Dust Slag Binders 156

References 173
Index 181

Preface

The problem of rational non-waste use of natural raw materials is one of the acute problems of modern civilization. According to UNESCO, 129 billion tons of ore fuel resources and raw materials for building materials are annually extracted from the earth worldwide as a result of human activities. For every inhabitant of the planet, there is an average of 20 tons of natural raw materials mined. The total area of destroyed land is about 20 million km^2, which is more than the total area used for agricultural needs. Intensive production of natural raw materials leads to an imbalance of the ecosystem and the need for special measures to stabilize and restore it.

With large volumes of extraction of natural raw materials, the degree of its usefulness, however, requires a significant increase. Annually, the world industry produces about 2100 million tons of solid wastes.

One of the most rational directions for the utilization of industrial wastes is their use as technogenic raw materials for various types of construction products. It is known that construction consumes about a third of the total mass of material production products, and material resources account for more than half of all the costs of construction and installation works.

The use of up to 40% of industrial waste in the production of building materials satisfies the need for raw materials, which reduces the cost of production of building materials from 10% to 30% compared to their production from natural raw materials, while saving capital investment of 35% to 50%.

The widespread introduction of superplasticizing additives in concrete and mortar technologies has opened up new radical ways of resource conservation in their production, including the use of technogenic raw materials based on various industrial wastes.

The traditional mineral additives that are introduced into Portland cement during its production are blast furnace granulated slag and fly ash. Coal fly ash and slag also are introduced directly into concrete and mortar mixtures successfully. The effectiveness of these mineral additives in cement systems has been proven by many years of practice and numerous scientific studies. At the same time, the introduction of blast furnace slag and ash into cement, concrete, and mortar mixtures in compositions with additives of superplasticizers (SP) has opened the possibility of a significant increase in their potential, and the production of materials with high physical and mechanical properties. The use of modern additives of superplasticizer has significantly revised the knowledge of the properties of cement concretes and mortars with a high content of slag and ash.

This monograph presents the results of studies of high-tech concretes based on composite cements, concretes based on low-clinker cements, effective modified mortars based on ashes, and slag dry building mixtures. Methods of designing concrete and mortar compositions with given properties, based on composite cements with the addition of polyfunctional modifiers, are given.

A significant place in the monograph is given to sulfate-fluoride and sulfate-fluoride-alkaline activation of low clinker slag Portland cement, the properties of concrete, and mortars based on activated binders.

The monograph also presents the results of studies that open up a new direction in the use of dust from clinker kilns in compositions with blast furnace granulated slag and Portland cement to obtain composite binders and mortars based on them.

A feature of the studies, the results of which are given in the monograph, is the widespread use of mathematical modeling based on experimental statistical methods. A large complex of mathematical models obtained in the work allows establishment of quantitative relationships between the properties of the studied materials and the main technological factors, prediction of the properties of concrete and mortars, and attainment of optimal technological decisions.

The studies, the results of which are presented in the monograph, were performed at the Department of Technology of Building Products and Materials Science of the National University of Water and Environmental Engineering (Ukraine).

The authors thank engineers Nichaeva L.I., Matsko L.A., and Kyts A.V. for their technical assistance in preparing the monograph for publication.

Authors

Leonid Dvorkin, Full Professor and Head of Department, National University of Water and Environmental Engineering (Ukraine). He is the author of many books and tutorials devoted to problems of concrete and binding technology. He has published many papers and presented lectures at national and international conferences on construction materials.

Vadim Zhitkovsky, Associate Professor, National University of Water and Environmental Engineering (Ukraine). He has published four monographs, six textbooks, and about 100 papers in scientific journals, edited books, and conference proceedings.

Mohammed Sonebi, Senior Lecturer and Director of Research of Environment Change and Resilience Cluster, Queen's University of Belfast. He authored/co-authored more than 220 peer-review journal and conference papers and 25 books/chapters with high citation ratings.

Vitaliy Marchuk, Senior Lecturer, National University of Water and Environmental Engineering (Ukraine). His research interests are technology of concrete, mortars and dry mixes, concrete mix use of industrial waste, and industrial products in the production of building materials.

Yurii Stepasiuk, Senior Lecturer, National University of Water and Environmental Engineering (Ukraine). His research interests are technology of concrete, concrete and mortars made based on low clinker slag Portland cements, and design of concrete with a set of specified properties.

1 Composite Ash-Containing Cements and Effective Concrete Based on Them

1.1 FLY ASH—AN ACTIVE FILLER OF CEMENT SYSTEMS

The influence of fly ash on the properties of concrete and mortars: An effective direction of reducing the consumption of binders and regulating the construction and technical properties of concretes and mortars is the introduction of active mineral additives (*active fillers*).

One of the effective active fillers of cement systems, as shown by numerous studies confirmed by practical experience, is fly ash [1–12]. Ash actively affects all stages of hydration and the formation of the structure of composite building materials, that is, a sequential transition from a coagulation structure to the formation of a spatial crystalline framework.

Fly ash, due to its glassy aluminosilicate phase, has pozzolanic activity and chemically interacts with $Ca(OH)_2$, which is released during the hydrolysis of clinker cement minerals. The introduction of fly ash into cement-water systems not only increases the volume of hydrated neoplasms, but also accelerates the hydrolysis process and increases the degree of cement hydration [5,6], which, ultimately, positively affects the strength of cement stone and mortars based on it.

Having a high specific surface, in addition to direct chemical interaction with cement, fly ash actively affects the physicochemical processes at the surface of the cement stone—filler interface.

The *activity of the ash* depends on the content of the vitreous phase. A close relationship has been established between the strength of the mortar containing ash and the calculated specific surface area of the vitreous phase. Glass in ashes [5] can be considered as a material containing amorphites—formations similar in composition and structure to the corresponding crystalline phases, but with a very high specific surface and disordered alumina-siliceous interlayers between them. The ability of the vitreous phase to hydrate and hydrolyze can be explained by the sub-microstructure and high permeability of amorphites due to the presence of cavities between ionic groups. The activity of the vitreous phase is determined by the ratio of alumina and silica. The larger the ratio is, the easier the process of hydration of fly ash is in an alkaline and sulfate-alkaline environment. In a neutral environment, the process of hydration of fly ash is stable. The hydraulic activity of calcium-aluminosilicate glass, which is contained in the ash, is positively affected by impurities of magnesium, iron, and some other elements [4].

A certain hydraulic activity in the ash, along with the vitreous phase, has a dehydrated and amorphized clay substance. Its activity depends on the mineralogical composition of clays, which are included in the mineral part of the fuel. With an increase in the content of amorphized clay substance in the ash, its water demand increases [13].

Important indicators of the quality of the ash that determine its activity are dispersion and particle size distribution. Numerous studies show that there is no direct relationship between these two indicators. Ash is characterized by a significant content of particles having small closed pores. They are the result of swelling of the molten mineral mass with gases released during the dehydration of clay minerals, dissociation of particles of limestone, gypsum, and organic substances. Pores can reach 60 % of the ash particles. The high content of micropores in the ash also determines the high value of its actual specific surface, the measurements of which, carried out by nitrogen adsorption [14], showed that it is an order of magnitude higher than the specific surface of cement. Its high specific surface area is associated with its properties such as adsorption capacity, hygroscopicity, and hydraulic activity.

In hydration systems, *"cramped conditions"* noticeably affect the physical interaction of particles [14,15], which are characterized by a sharp increase in the volume concentration of the solid phase and the transition of part of the bulk water to film. The creation of cramped conditions allows technologically justified to reduce the water demand of the mortar mixture. The use of ash as an active filler with the constancy of other parameters of the mixture contributes to the formation of such conditions at the coagulation stage of structure formation.

A consequence of the active influence of ash on the hydration and formation of the structure of cement stone is a significant change in the properties of concrete and mortar mixtures. The introduction of ash into the mortar mixture, unlike other fillers, does not worsen, but improves the mobility of the mixture. Already the first researchers found that the dependence of the mobility of the mixture on the ash content is extreme, and its optimal content should be no more than 30% of the binder mass [16]. The plasticizing effect of ash is influenced by the shape, state of the surface of the particles, and their dispersion. The glassy surface of the ash particles reduces internal friction and reduces the viscosity of the mortar.

Many researchers believe that spherical ash particles can be considered as solid "ball bearings" in a mixture of binder, water, and aggregates, similar to how emulsified air bubbles, when using air-entraining additives, have a plasticizing effect on the mortar mixture, acting as kinds of air ball bearings. Larger fractions contain more unburned carbon particles, which have increased water absorption and irregularly shaped particles [4]. Therefore, the water demand when using low-dispersion ashes significantly increases.

An increase in the dispersion of ashes and a decrease in their water demand can be achieved by ash selecting of electro-static precipitators from the last fields or by grinding, which destroys the organomineral aggregates. The decrease in ash water demand during grinding is explained by a decrease in the amount of capillary water held by aggregated particles. This decrease leads to a more significant effect than the increase in the amount of adsorbed film water with increasing specific surface area.

The introduction of ash helps to reduce the water separation of the mortar mixture [14]. The plasticizing and water-retaining ability of ash determines the promising use of it primarily in masonry solutions. Mortar mixtures with an optimal ash addition have a rather high "viability" and are suitable for transportation over long distances [4].

The effect of ash on the strength of concrete and mortars depends on its properties and dispersion, content and chemical and mineralogical composition of cement, age and processing conditions. To assess the effect of ash on the strength of concrete and mortars, the concept of its *"cementing efficiency"* is introduced, which is characterized by a coefficient of $K_{c.e}$ [4]. When predicting the strength of concrete and mortars, it is proposed to find the *modified cement-water ratio* by the formula

$$(C/W)_m = \frac{C + \kappa_{c.e} m_a}{W}, \qquad (1.1)$$

where C is the consumption of cement in ash contained concrete, W is the water consumption, and m_a is the mass fraction of ash in a mixed binder.

According to the data in [16], neither the carbon content nor the specific surface can be used to estimate the $K_{c.e}$ of the ash. Ash, which has the lowest values of specific surface area and different carbon contents, showed the highest values of the coefficient $K_{c.e}$. To determine the values of $K_{c.e}$, determine the strength of the material at various with the addition of ash and determine $(C/W)_m$, etc. In [4], it was found that when using Portland cement and the ash of the district power station, the value of the coefficient $K_{c.e}$ varies from 0.2 to 0.4.

Having determined the values of $(C/W)_m$ and having set the optimum ash content with the known $K_{c.e}$, it is possible to determine the necessary C/W of ash-containing mortars or concrete and to design their compositions.

Most researchers note the positive effect of increasing the dispersion of ash on the strength of cement-ash mixtures, especially at an early age. It has been established that the activity of ash increases significantly when the size of its particles is 5 to 30 microns [5,17,18].

The most favorable effect of the ash additive on the strength of composite materials is observed with a relatively low consumption of binders.

To achieve high strength ash-containing cement composites, the chemical and mineralogical composition of clinker is of decisive importance. At an early age, an increase in the strength is promoted by an increased content of alkalis in the clinker, which accelerate the chemical interaction of ash and cement [19]. At a later age, for the manifestation of the pozzolanic reaction of ash, the best are cements with a high content of alite, which upon hydrolysis form a significant amount of $Ca(OH)_2$. It is advisable to add ash to materials on Portland cement with a high content of minerals—silicates [4].

Ash not only increases the cohesive and adhesive strength of the cement matrix, but also, which is very important, reduces the voidness of the aggregate [14,40].

With a certain lag in the strength of ash-containing mortars and concrete in the early stages, many authors have shown that in the period of 28 to 180 days the growth rate of their compressive strength is approximately the same or higher than without the addition of ash.

In some works, prolonged hardening increases intensively the strength of ash-containing mortars under tension and bending. Samples in the form of rods and bars cut from an experimental mortar masonry showed bending strength of ash-containing solutions after 3 months was 80%, and after 10 years was 150%, of the strength of the control samples. In mortars with ash, as well as with other active mineral additives, a higher ratio of tensile strength to compressive strength is characteristic [5].

Many studies contain conflicting data on the modulus of elasticity, creep, and shrinkage during drying of ash-containing mortars and concrete [4,5,20]. This conflict can be largely explained by the great influence that the strength of hardened materials and the density of the aggregate have on these indicators. For example, in [19] it is stated that the replacement of a part of cement with ash leads to a decrease in the shrinkage deformation of concrete. The decrease in shrinkage is explained by the fact that ash adsorbs soluble alkalis from cement and forms stable, insoluble aluminosilicates. In [20], it was shown that the shrinkage during drying of prisms from concrete without additives and concrete with the addition of ash is basically the same after 400 days, although in the initial period there is a tendency of growth with the introduction of ash.

Ash helps to increase the sulfate resistance of cement mortars and concretes in the same way as other active mineral additives.

In the experiments of V. V. Stolnikov [21], the cement stability coefficient after 6 months of exposure to a sulfate solution was 0.33, while when ash was added it exceeded 1, which indicates the complete absence of corrosion of the samples. The best results were noted with the introduction of ash with the highest content of $SiO_2 + Al_2O_3$, that is, the most acidic in chemical composition.

According to the recommendations [4], when using reactive aggregates containing opal, chalcedony, silicon shales, volcanic tuffs, etc., in the mortar mixture, ash can be used only if the total content of alkaline oxides in the binder in terms of Na_2O will be no more than 0.6% by weight. At the same time, in [16], the recommendation of the upper permissible limit of the possible total content of alkaline oxides in a cement-ash binder is 1.5%.

Like other active mineral additives, fly ash can cause a slight decrease in frost resistance of mortars and concrete. It was found that when the dosage of ash is 30% to 40% of the mass of the mixed binder, a sharp increase in capillary suction and water absorption is observed, especially in the initial test period. This increase in capillary suction and water absorption indicates an increase in the absolute volume of open capillary pores. The filling of open capillary pores with water occurs under the action of hydrostatic pressure, while the total porosity, although growing, is much less. Studies have shown that when replacing cement with ash, the volume of contraction pores is significantly reduced. As the ash content increases, the ratio between capillary and contraction porosity changes. Capillary water absorption with the addition of ash to the cement increases by about 10% to 20% for every 10% of the ash additive.

The degree of decrease in frost resistance of mortars and concrete with the introduction of ash in them is different and depends on its characteristics. Significant differences in the basic physical and mechanical properties of cement-ash compositions, including frost resistance, are caused by heterogeneity of the composition and properties of ash.

A significant increase in frost resistance of ash-containing mortars and concrete is achieved by the introduction of surfactant additives.

Many researchers have shown that the effectiveness of the addition of fly ash in cement mortars and concrete significantly increases with the content of surfactant additives. As shown in [22], the introduction of a surfactant can be considered as one of the methods for activating the ash filler in mortars and concrete. When choosing a surfactant, the chemical nature of both the filler and the binder should be considered.

Much current experimental data demonstrates the effectiveness of joint introduction into the mortar and concrete mixtures of fly ash and superplasticizers (SP) [23,24]. According to the mechanism of action, most SPs can be attributed to the second and third groups of surfactants according to the classification of P. A. Rebinder [25]. Common to them is the ability to peptization (deflocculation) the aggregated cement flocs. The water immobilized in the flocs is released and helps to dilute the cement paste. Compared with traditional plasticizers, for example, the lignosulfanate type, SPs have a longer hydrocarbon chain and, correspondingly, a higher adsorption capacity [26]. A significant influence on the mechanism of action of SP in concrete and mortar mixtures has colloidal chemical phenomena at the phase boundary and the magnitude of the zeta potential [27–29].

By now there are several recommendations for the activation of fillers by modifying the surface with various chemicals, including surfactants, halide-containing organosilicon substances, as well as by mechanical and radiation treatment of binders and fillers. Most of the methods that increase the surface activity of binders are acceptable for fillers including ash. The classification of activation methods for fillers and the effect of activation on the mechanical properties of cement concrete are given in [4].

The main investigated methods for activating fly ash include grinding (*mechanical method*) and surface modification of particles with mineral or organic additives (*chemical method*). A combination of both methods (*mechanochemical method*) is considered promising [30].

The essence of the mechanical method is to increase the reactivity of powders by opening new active grain surfaces, changing the crystalline structure of minerals, forming a large number of unsaturated valence bonds, and during deep grinding and their amorphization. In the most common cases, the surface energy of the filler E_f is lower than the surface energy of the binder grains E_b ($E_b/E_f > 1$). The relative ratio between the particle size of the filler d_f and the binder grains d_b for these cases is recommended in the range $d_f/d_b = 3$ to 10 [29]. These recommendations are developed based on ideas about the formation of clusters and the phased organization of the structure of composite material. In cases where the surface activity of the filler is higher than the surface activity of the binder, the optimal particle size of the filler is recommended equal to the size of the cement grains [4]. A further decrease in the diameter of the filler particles increases the likelihood of their association into their own cluster structures, which leads to the development of internal interfaces and crack formation. The size of the filler-binder phase interface in a unit volume can be characterized by a parameter proportional to the product of the degree of filling, dispersion, and surface activity of the filler.

The feasibility of activating the filler by modifying its surface with surfactant additives follows from the Dupre-Young equation [31], which relates the work of adhesion W_{ad} to the surface energy of a solid:

$$W_{ad} = v_s - v_s^*(m + \cos\theta), \quad (1.2)$$

where v_s is the surface energy of a solid, v_s^* is the free surface energy of a solid in the atmosphere of vapors and gases, $m = v^*_l v_l > 1$ (v_l^* is the surface tension of the liquid oriented under the influence of the force field of the solid surface, v_l is the surface tension of the wetting liquid), and θ is the angle of wetting. From the equation it follows that to achieve high adhesive strength it is important to ensure the necessary wettability of the filler with a binder and to reduce interfacial surface energy, which is achieved by surfactant processing of the filler.

The decrease in interfacial surface energy when creating an adsorption-active medium is determined from the equation

$$\Delta v_{s,l} = KT \int_0^c n_s(c) d\ln c \quad (1.3)$$

where $\Delta v_{s,l}$ is the difference in interfacial surface energy without a surfactant and in the presence of a surfactant with a concentration c, n_s is the adsorption value determined by the number of surfactant molecules adsorbed on 1 cm² of the interface, K is the Boltzmann constant, and T is the absolute temperature in °C.

A necessary condition for the effectiveness of surfactants is the ability of chemisorption with the surface of the filler particles. In the general case, cationic surfactants are recommended [30] for mineral fillers of the acid type, and anionic surfactants for mineral fillers of the basic type. The influence of the adsorption-active medium on the value of $\Delta\sigma_{s,l}$ increases with increasing dispersion of the filler and its concentration, which is associated with an increase in the interfacial surface and, accordingly, excess surface energy.

When choosing a method for activating ash as mineral filler, it is necessary to take into account, along with the effect of increasing its activity, the manufacturability and energy intensity of the processing of the powder component, the possible productivity of the aggregates, and controllability and stability of the achieved parameters of the activated material. The most developed mechanical and mechanochemical methods of activating fillers, based on the regrinding of powdered components, have several disadvantages. The main ones are high energy intensity and the need for a precast concrete plant to install a rather complicated production line, including a grinding unit, dust removal, and other equipment. These shortcomings did not allow, in particular, realization of the well-known idea of activating dispersed ashes by grinding them, although the technical and economic effect achieved in this case is beyond doubt [30].

A more accessible way is to increase the activity of fillers in activator mixers, thus creating high velocity gradients and turbulent flows, due to which the energy state of the surface of the powders change and their activity increases. This method finds application, albeit limited, for the activation of cements and filled cement paste, where the achieved effect of deflocculation is especially important [32]. Insufficient ash and ash-slag reliability of mixer-activators, and instability of the achieved results hinder the spread of this method.

An effective way seems to be the activation of mineral fillers by combining them with such highly active silica additives as silica fume, metakaolin, etc. [4,5].

This book presents the results of the efficiency of activation of fly ash and ash-slag fillers by a polyfunctional modifier (PFM), which includes SPs and other chemical additives.

1.2 POZZOLANIC ACTIVITY, HYDRATION FEATURE, AND RHEOLOGICAL PROPERTIES OF CEMENT-ASH PASTES WITH ADDITIVES MODIFIERS

1.2.1 Pozzolanic Activity

Pozzolanic activity of the fly ash largely depends on dispersion, chemical and phase composition, and especially the content of the vitreous phase [5,33–38].

The highest results were obtained with the use of fine ash with low unburnt coal content [3]. When filling the aqueous layers of the curing system with pozzolanic reaction products, the strength of the concrete is increased due to the formation of strong bonds between the ash particles and the hydration products of the cement. However, until these layers are filled, the pozzolanic reaction has no significant effect on the strength gain [14].

The study of the features of composite cement (CC) hydration processes with the addition of PFMs, which included SP, grinder, and curing accelerator, determined the pozzolanic activity of fly ash at different values of the specific surface area (S_{sp}) (according to the absorption of CaO). For this purpose, the pre-heated and thermostated at 80°C saturated lime solution and the investigated ash powders were mixed and after a certain holding time were measured by titrating the concentration of $Ca(OH)_2$. In experiments, the fly ash with chemical composition was used: ($SiO_2 + Al_2O_3 + Fe_2O_3$), 85.8%; SO_3, 2.3%; CaO_v, 2.8%; MgO, 2.0%; ($Na_2O + K_2O$), 1.2%; and loss on ignition, 5.1%.

The mixture with the ash used blast furnace slag with the following chemical composition: SiO_2, 39.1%–39.9%; Al_2O_3, 6.33%–6.65%; Fe_2O_3, 0.11%–0.18%; CaO, 46.8%–47.4%; MgO, 3.08%–3.22%; MnO, 1.14%–1.21%; SO_3, 1.69%–1.76%.

The results of studies of the pozzolanic activity of fly ash and its compositions with slag at different dispersion in the presence of PFM are given in Table 1.1 and in Figure 1.1.

TABLE 1.1
Pozzolanic Activity of Fly Ash and Ash-Slag Compositions

No s/n	Name	CaO Absorption (mg/g)		
		7 Days	28 Days	60 Days
1	Fly ash (S_{sp} = 350 m²/kg)	15	52	78
2	Fly ash (S_{sp} = 350 m²/kg)	18	65	97.5
3	Fly ash (S_{sp} = 350 m²/kg)	25	90	135
4	Fly ash + PFM_1 [a] (S_{sp} = 450 m²/kg)	19	67	100.5
5	Fly ash + PFM_2 [b]	18	63	94.5
6	Fly ash + PFM_3 [c]	18	67	100.5
7	Fly ash + slag (ash:slag, 3:1, S_{sp} = 450 m²/kg)	18	66	99
8	Fly ash + slag (ash:slag, 1:1, S_{sp} = 450 m²/kg)	19	68	102
9	Fly ash + slag + PFM_1 (ash:slag, 3:1, S_{sp} = 450 m²/kg)	25	71	106.5
10	Fly ash + slag + PFM_1 (ash:slag, 1:1, S_{sp} = 450 m²/kg)	27	73	109.5
11	Fly ash + slag + PFM_2 (ash:slag, 3:1, S_{sp} = 450 m²/kg)	21	68	102
12	Fly ash + slag + PFM_2 (ash:slag, 1:1, S_{sp} = 450 m²/kg)	23	70	105
13	Fly ash + slag + PFM_3 (ash:slag, 3:1, S_{sp} = 450 m²/kg)	26	72	108
14	Fly ash + slag + PFM_3 (ash:slag, 1:1, S_{sp} = 450 m²/kg)	27	74	111

[a] Polycarboxylate type SP Sika VC 225 + propylene glycol (PG) (PFM_1 = 0.7%).
[b] SP naphthalene formaldehyde type SP-1 + PG (PFM_2 = 1.5%).
[c] Sika VC 225 + PG + sodium sulfate (PFM_3 = 1.7%).

According to the graphical dependence (Figure 1.1), obtained according to Table 1.1, the kinetics of CaO absorption by ash and its blast furnace compositions can be roughly divided into two periods: the first corresponding to a sharp initial change in $Ca(OH)_2$ concentration and the second characterized by a slower process rate. Such separation of the process of absorption of lime by colloidal silica was noted earlier in [5]. According to this paper, the first period of this process is the adsorption of $Ca(OH)_2$ on SiO_2 particles. The amount of adsorbed $Ca(OH)_2$ approximately corresponds to the layer of Ca^{2+} ions on the surface of SiO_2 particles. Further long-term decrease in the concentration of calcium hydroxide is due to the precipitation of a solution of calcium hydrosilicate, which was formed as a result of a chemical reaction.

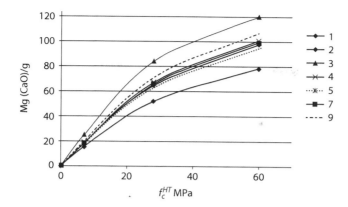

FIGURE 1.1 Kinetics of CaO absorption by fly ash and fly ash compositions + blast furnace slag (designation according to Table 1.1).

Pozzolanic ash reaction according to several researchers [2,14] also begins with adsorption of calcium hydroxide on its surface. The thin boundary layers resulting from this adsorption serve as conductors of calcium ions under the action of which gradual erosion of the surface of the ash particles develops. The erosion forms recesses where the products of the pozzolanic reaction settle.

The introduction of PFM, as can be seen in Table 1.1, does not significantly affect the absolute magnitude of pozzolanic activity, but at the same time it changes the kinetics of absorption of lime ash. When using PFM_1 and PMF_3, a higher rate of CaO uptake is observed compared to PMF_2. Increasing the dispersion also increases the amount of CaO absorbed by the fly ash and the fly ash compositions + blast furnace slag.

1.2.2 Features of Hydration

The peculiarities of the hydration processes of fine-ground cement-ash pastes with the addition of PFM were studied by a chemical method. The degree of hydration of cement α [39] was determined by the formula:

$$\alpha = h/W, \tag{1.4}$$

where h is the amount of hydrated water (i.e., that does not evaporate at 105°C) attached to 1 g of cement after a certain time, W is the amount of non-volatile water attached to 1 g of cement with almost complete hydration.

For this purpose, samples were made of ash-containing CCs with a water-binder ratio of 0.3, which were cured under normal conditions for a specified period, then crushed, treated with acetone to remove free water, and subjected to calcination to determine the amount of chemically bound water. The numerical values of the degree of hydration of pastes based on CC with the addition of PFM are given in Table 1.2.

From Tables 1.1 and 1.2 it follows that PFM additives have different effects on the process of CC hydration. If the PFM_2 additive containing the

TABLE 1.2
The Degree of Hydration of Ash-Containing CCs with PFM Additives

Kind PFM	The Content of Fly Ash in the CC (%)	Contents PFM (%)	The Degree of Hydration		
			3 Days	7 Days	28 Days
PFM$_1$	30	1.0	0.22	0.39	0.52
		0.7	0.22	0.39	0.50
		0.4	0.21	0.37	0.49
	40	1.0	0.21	0.37	0.50
		0.7	0.20	0.37	0.48
		0.4	0.20	0.35	0.47
	50	1.0	0.18	0.35	0.47
		0.7	0.17	0.35	0.45
		0.4	0.17	0.34	0.44
PFM$_2$	30	2.0	0.16	0.32	0.47
		1.5	0.17	0.32	0.48
		1.0	0.17	0.34	0.48
	40	2.0	0.14	0.30	0.45
		1.5	0.15	0.30	0.45
		1.0	0.16	0.32	0.47
	50	2.0	0.14	0.29	0.41
		1.5	0.14	0.30	0.42
		1.0	0.15	0.31	0.43
PFM$_3$	30	2.0	0.24	0.38	0.54
		1.7	0.22	0.36	0.52
		1.4	0.22	0.34	0.51
	40	2.0	0.23	0.38	0.62
		1.7	0.21	0.37	0.60
		1.4	0.21	0.36	0.60
	50	2.0	0.20	0.36	0.48
		1.7	0.19	0.35	0.47
		1.4	0.17	0.34	0.46

Note: $S_{sp} = 450$ m^2/kg; PFM$_1$ (polycarboxylate SP) Sika VC 225 + PG; PFM$_2$ (naphthalene formaldehyde SP-1) + PG; PFM$_3$, Sika VC 225 + sodium sulfate + PG.

naphthalene-formaldehyde-type SP reduces the hydration degree of the cement in 3 to 7 days, then other modifiers increase it by 15% to 20%. At 28 days, the additives have little or no effect on the degree of CC hydration.

The effect of PFM$_1$ and PFM$_3$ modifiers on CC hydration is not fundamentally different. Some acceleration of these processes in the presence of the PFM$_3$ modifier at the initial stage of hydration is practically counterbalanced by increasing the duration of curing. Therefore, it can be considered that the use of modifiers containing accelerators of setting and hardening is only appropriate if an accelerated initial structure formation is required.

1.2.3 Rheological Properties

The complex of construction and technical properties of composite ash-containing cements with PFM additives is largely determined by their rheological properties, in particular the *effective viscosity* of cement-ash paste. In the course of these experiments, mathematical planning was applied. For this purpose, a three-level three-factor plan B_3 was implemented [40]. The conditions for planning the experiments are given in Table 1.3.

When investigating the viscosity of cement-ash paste based on CC with PFM additives a rotational viscometer was used—the main feature of which is the ability to take into account bottom and end effects. An important advantage of rotational rheometry is the ability to implement with it almost unlimited shear deformations with a given speed.

Pastes with W/C = 0.3 were investigated after mixing when determining the rheological properties after 5 minutes. Paste viscosity (η) was determined by the formula:

$$\eta = k \frac{P - P_o}{N} \tau, \tag{1.5}$$

where k is the device constant, P is the weight of cargo, N is the speed, τ is the rotation time, and P_o is the idle load (P_o = 12.4 g).

The speed of the cylinder varied from 27 to 430 c^{-1}. The results of experiments to determine the effective viscosity of cement-ash paste based on CC with PFM additives are given in Table 1.4.

The analysis of the received data given in Table 1.4 showed that in fine-grained cement-ash paste the greatest decrease in effective viscosity is achieved by increasing the content of PFM and correspondingly increasing the content of the SP (Figure 1.2). The highest viscosity reduction is achieved with the maximum number of PFMs. The specific surface of the binder has less effect on the viscosity, increasing it by 35 to 45% at a dispersion of 590 m²/kg compared to the dispersion of the CC at 330 m²/kg. Among the test modifiers, ceteris paribus (dispersion and ash content), the viscosity of the PFM cement-ash paste containing the polycarboxylate type SP

TABLE 1.3
Experimental Planning Conditions

Factors		Levels of Variation of Factors			Variation Interval
Natural	Coded	−1	0	+1	
The content of fly ash in the CC (%)	x_1	30	40	50	10
Specific surface CC, S_{sp} (m²/kg)	x_2	330	460	590	130
Contents PFM$_1$/PFM$_2$/PFM$_3$ in CC (%)	$x_{3(1)}$	0.4	0.7	1	0.3
	$x_{3(2)}$	1.0	1.5	2	0.5
	$x_{3(3)}$	1.4	1.7	2	0.3

TABLE 1.4
The Results of Experiments on the Determination of the Effective Viscosity of Cement-Ash Paste with PFM Additives

	The Composition of Cement-Ash Paste			Viscosity, 10^{-1}, Pa·c		
No.	Ash (%)	S_{sp} (m²/kg)	$PFM_1/PFM_2/PFM_3$ (%)	PFM_1	PFM_2	PFM_3
1	50	590	1.0/2.0/2.0	41.9	43.4	47.2
2	50	590	0.4/1.0/1.4	100.3	114.7	126.5
3	50	330	1.0/2.0/2.0	12.8	15.3	16.5
4	50	330	0.4/1.0/1.4	82.6	95.5	104.5
5	30	590	1.0/2.0/2.0	46.1	48.5	52.3
6	30	590	0.4/1.0/1.4	103.3	118.7	132
7	30	330	1.0/2.0/2.0	16.8	19.3	21.5
8	30	330	0.4/1.0/1.4	84.2	98.4	110
9	50	460	0.7/1.5/1.7	59.8	78	85.8
10	30	460	0.7/1.5/1.7	56.1	70.8	77
11	40	590	0.7/1.5/1.7	63.5	86.9	96.8
12	40	330	0.7/1.5/1.7	48.2	53.1	58.3
13	40	460	1.0/2.0/2.0	39.1	42.8	46.2
14	40	460	0.4/1.0/1.4	99.2	113.3	123.2
15	40	460	0.7/1.5/1.7	57.2	74.6	81.4
16	40	460	0.7/1.5/1.7	58.2	76	83.6
17	40	460	0.7/1.5/1.7	57.6	74.9	81.4

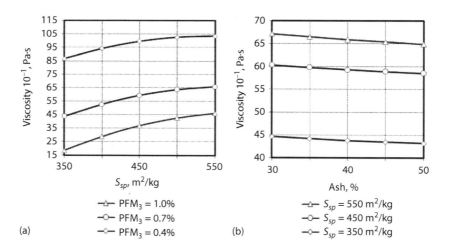

FIGURE 1.2 The dependence of the effective viscosity of CC-based pastes on technological factors using (a) PFM_3 (0.4–1%) and (b) PFM_1 (0.7%).

decreases the most. Fly ash at constant W/C, which increases the total concentration in the paste dispersed phase and reduces the water-binder ratio at W/C = const, naturally causes an increase of viscosity. However, with the introduction of PFM additives, this effect is largely smoothed out, especially in the presence of the Sika VC 225 SP.

Statistical processing of experimental data allowed obtaining mathematical models of effective viscosity of cement-ash paste with PFM additives in the form of polynomial regression equations with coded factor values:

When using PFM_1:

$$\eta_1 = 59.3 - 0.91x_1 + 11.1x_2 - 31.3x_3 - 0.2x_1x_2 + 1.13x_1x_3 \\ + 2.7x_2x_3 - 2.38x_1^2 - 4.48x_2^2 - 0.45x_3^2. \quad (1.6)$$

When using PFM_2:

$$\eta_2 = 76 - 0.88x_1 + 13.6x_2 - 37.13x_3 - 0.28x_1x_2 \\ - 0.285x_1x_3 + 2.23x_2x_3 - 1.96x_1^2 - 6.36x_2^2 + 1.96x_3^2. \quad (1.7)$$

When using PFM_3:

$$\eta_3 = 83 - 1.23x_1 + 14.4x_2 - 41.25x_3 - 0.012x_1x_2 + 1.13x_1x_3 \\ + 2.19x_2x_3 - 2.01x_1^2 - 5.86x_2^2 + 1.29x_3^2. \quad (1.8)$$

The analysis of the obtained models shows that the dominant influence on the cement paste viscosity based on ash-containing CCs is due to the addition of SP. The influence of polycarboxylate SPs with a constant ash content and an S_{sp} of 15% to 25% higher than when using naphthalene-formaldehyde SP. Increasing the fineness of grinding cement from 330 to 590 m²/kg increases the viscosity of the paste with a constant content of the SP additive 1.1 to 1.2 times. The increase in the CC ash additive is not significantly reflected in the presence of PFM on the viscosity of the cement-ash paste.

To ensure enough "viability" of mortar mixtures, the nature of change in viscosity of the cement-water matrix over time is crucial. The change in viscosity over time was performed on typical cement and cement-ash paste with PFM additives of different composition at 20°C ± 2°C. The results of experiments are shown in Figure 1.3.

Figure 1.3 implies that pastes containing PFM_2 must have a higher effective viscosity when held. Pastes with PFM_1 and PFM_3 additives have a lower viscosity in the early term. The presence of PFM_3 curing accelerator at low initial viscosity results in a noticeable increase in viscosity after 35 minutes. The introduction of PFM additives at the optimum content compensates for the increase in the viscosity of the cement-ash paste with an increase in the content of fly ash and specific surface area. The highest viability of the pastes is ensured by the introduction of PFM containing polycarboxylate type SP.

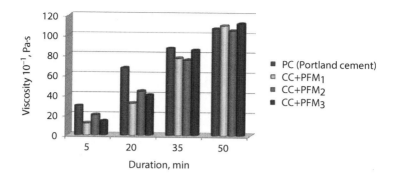

FIGURE 1.3 The influence of the duration of keeping the type and amount of PFM on the paste viscosity based on CC (ash = 40%; PFM$_1$, 0.7%; PFM$_2$, 1.5%; PFM$_3$, 1.7%; and S_{sp} = 450 m²/kg).

1.3 STRUCTURE FORMATION AND PHASE COMPOSITION OF ASH-CONTAINING COMPOSITE CONCRETE

1.3.1 THE INITIAL STRUCTURE FORMATION

In cement systems, the process of structure formation is divided into three main periods:

1. The induction period: this takes place directly in the process of mixing the binder with water.
2. The period of initial or coagulation structure formation: this coincides practically with the setting process.
3. The final period is the period of crystallization structure formation.

According to [14], the limits of these stages are determined by the characteristic time τ_1, τ_2, τ_3—points of a qualitative change in the structure of the cement system. Time τ_1 characterizes the conditional end of the induction period and the beginning of the setting process, which ends during τ_2. Then the primary, mainly aluminate, structural network begins to form. The crystallization growth of neoplasms of the main carriers of strength—silicate minerals—begins later and time τ_3 can be characterized.

An important structural and mechanical characteristic of the hardening of disperse systems is *plastic strength*. The value of plastic strength in the initial period of hardening of cement and cement-ash pastes based on CC was determined using a conical Plastomer with a division value 0.01 mm, while the depth of immersion of the cone was fixed with an indicator. The value of plastic strength was determined by the formula:

$$P_m = 0.096 \, P/h^2, \tag{1.9}$$

where P is the weight of the cone with the load, N and h are the immersion depths of the cone in millimeters.

Based on the obtained plastograms, the conclusion is that they have two characteristic sites. This conclusion is confirmed in the work of many researchers who think that the first plastogram section approximately corresponds to the setting time of pastes, and the inflection point corresponds to the end of setting [41]. O. E. Kalmykova and M. V. Mikhailov in their studies showed that plastic strength characterizes the ultimate shear stress of a disperse system at a certain point during solidification. Changing the plastic strength of cement pastes is an integral characteristic of the course of coagulation processes and crystallization structure formation. Moreover, if the first section of the plastogram characterizes mainly the formation of the thixotropic coagulation structure, then the second is the period of coagulation strengthening and the beginning of the formation of the crystallization structure. An analysis of the plastogram shows that for ash-containing CC with PFM additives, the characteristic period of coagulation structure formation is slightly increased. In cement-ash pastes with SP and hardening accelerator as a part of PFM, there is a more intensive increase in plastic strength in the second section of the plastogram.

Particles of fly ash form coagulation contacts with hydrated particles of cement. Under such conditions, for the existing structures, the contact strength ($f_{c.s}$) depends on several factors [4]:

$$f_{c.s} = \gamma \cdot f(F_r, \varphi, S_{sp}), \qquad (1.10)$$

where γ is the constant of chemical interaction, F_r is the resultant force of interaction between particles, φ is the degree of filling, and S_{sp} is the specific surface of the particles involved in the interaction.

The introduction of fly ash into the composition of CC in combination with PFM in this area leads to an increase in plastic strength, which corresponds to known data [42]. After passing 10 to 12 hours of hardening, the plastic strength of the modified pastes increases significantly and reaches 3.5 Pa when using PFM_3, 2.9 Pa when using PFM_1, and 2.5 Pa when using PFM_2. To achieve the same value of plastic strength for unmodified pastes requires a much more time.

The kinetics of the set of plastic strength (Figure 1.4) were compared with the kinetics of the velocity of ultrasonic waves passing through hardening pastes. In parallel with the determination of the propagation velocity of longitudinal ultrasonic waves, the change in electrical resistance was determined.

The results of experimental studies of the structural features of cement-ash pastes based on CC with PFM additives by the speed of transmission of longitudinal ultrasonic waves (V_{luw}, m/s) are shown in Figure 1.5.

As can be seen from these figures, the curves of the dependence of the speed of passage of ultrasonic waves through the cement paste in time contain sections, also characteristic of the stages of the formation of the structure: at the initial stage (induction period), V_{ulw} increases slightly; further, due to the formation of the crystallization frame of the aluminate components of the cement $dV_{ulw}/d\tau$, it becomes

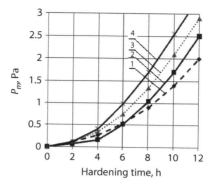

FIGURE 1.4 Kinetics of changes in the plastic strength of cement-ash pastes with PFM additives (ash = 40% by weight of CC): (1) cement-ash paste with Sika VC 225 = 0.7%, (S_{sp} = 330 m²/kg), (2) with the addition of PFM_2 (PFM_2 = 1.5%, S_{sp} = 460 m²/kg), (3) with the addition of PFM_1 (PFM_1 = 0.7%, S_{sp} = 460 m²/kg), and (4) with the addition of PFM_3 (PFM_3 = 1.7%, S_{sp} = 460 m²/kg).

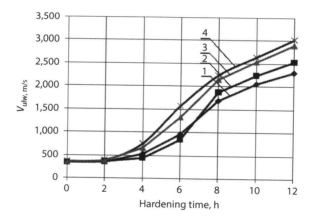

FIGURE 1.5 Change in ultrasound velocity in modified cement-ash pastes (ash = 40%): (1) cement-ash paste with Sika VC 225 = 0.7%; S_{sp} = 330 m²/kg, (2) with the addition of PFM_2 = 1.5%; S_{sp} = 460 m²/kg, (3) with the addition of PFM_1 = 0.7%; S_{sp} = 460 m²/kg, and (4) with the addition of PFM_3 = 1.7%; S_{sp} = 460 m²/kg.

maximum; with the beginning of crystallization of neoplasms of hydrosilicates, which are the main carriers of strength, the growth rate of V_{ulw} increases.

The initial period of formation of the coagulation structure on the ultrasound velocity curve is characterized by a horizontal section, the length of which coincides with the initial setting. Determining the initial setting by the value of plastic strength

Composite Ash-Containing Cements and Effective Concrete Based on Them 17

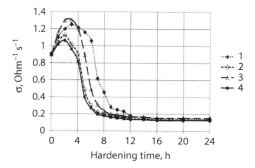

FIGURE 1.6 Change in the electrical conductivity of cement-ash pastes with PFM additives (ash = 40% by weight of CC): (1) cement-ash paste with SP Sika VC 225 = 0.7%; S_{sp} = 330 m²/kg, (2) with the addition of PFM_2 = 1.5%; S_{sp} = 460 m²/kg, (3) with the addition of PFM_1 = 0.7%; S_{sp} = 460 m²/kg, and (4) with the addition of PFM_3 = 1.7%; S_{sp} = 460 m²/kg.

is difficult due to its lower sensitivity to initial contacts between particles. The rate of structure formation can be related to the angle of inclination of the ultrasound velocity curve as well as of plastic strength to the abscissa axis.

With the kinetics of growth of plastic strength and the speed of ultrasound, the kinetics of the electrical conductivity of the pastes is proportionally changed—the inverse of the electrical resistivity (Figure 1.6). The change in the electrical resistivity of the samples (σ, $Ohm^{-1} \cdot m^{-1}$) was measured using a rheochord bridge at a current frequency of 50 Hz and a cell for plexiglass samples with an interval of 30 minutes.

Particles of cement when interacting with water form an ionic medium capable of conducting an electric current. The electrical conductivity of such a medium increases in proportion to the degree of dissociation of electrolytes (cement minerals in water) [43]. After a period of time (usually 0.5–1 hours), it reaches its maximum, after which the value of electrical conductivity changes insignificantly over time, and then it decreases (Figure 1.6).

As noted in [14,15], during the first 40 minutes the processes of hydrolysis of minerals of Portland cement clinker (mainly C_3A and C_4AF) and the formation of a saturated solution of calcium hydroxide intensively occur, an increase in the concentration of conductive ions per unit time. The maximum electrical conductivity reflects the beginning of the induction period [14], during which intensive adsorption of water by minerals, the formation of the coagulation structure of cement pastes, and the formation of crystal nuclei occur. Thus, the analysis of the studies performed, and the nature of the change in plastic strength, the speed of ultrasound, and the electrical conductivity of the pastes on the studied CCs indicate that increasing the fineness of grinding and the introduction of PFM accelerates the structure formation processes in the entire range of the content of fly ash in the CC.

1.3.2 Phase Composition and Porosity of Cement Stone

Peculiarities of the phase composition and microstructure of cement stone based on ash-containing CCs with PFM additives were studied using X-ray diffraction analysis and electron microscopy.

X-ray studies were carried out using a DRON-3 diffractometer.

Analysis of diffractograms of hydrated ash CC with PFM additives of various compositions (Figure 1.7) indicates that their hydration processes are active with the formation of the main hydrated phases.

Non-hydrated Portland cement is characterized by intense lines of alite and belite phases with interplanar distances (d/n = 0.277; 0.260; 0.218 nm). During hydration of Portland cement (PC I), portlandite lines (d/n = 0.263; 0.491 nm) and a low intensity ettringite line (d/n = 0.973; 0.561 nm) appear on the X-ray diffraction pattern.

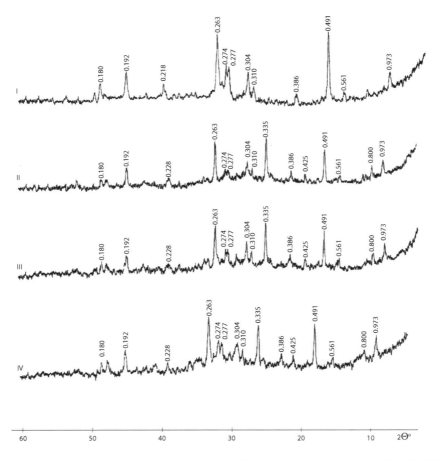

FIGURE 1.7 Diffraction patterns of cement samples of various compositions: (I) PC-I-500, hydrated at 28 days; (II) CC (Portland concrete (PC), 50%; ash, 50%; and S_{sp} = 460 m²/kg); (III) CC (PC, 50%; ash—50%; Sika VC 225, 0.7% S_{sp} = 460 m²/kg); (IV) CC (PC, 50%; ash—37.5%, slag—12.5%, Sika—0.7%, S_{sp} = 460 m²/kg).

The introduction of 50% (by weight) active mineral additives (blast furnace granulated slag and fly ash) into the CC is accompanied by a regular decrease in the intensity of the lines of the main clinker minerals (C_3A, C_3S and β-C_2S), as well as calcium hydroxide $Ca(OH)_2$. The presence in the binder of blast furnace granulated slag causes the appearance of a sufficiently large amount of active amorphous vitreous phase.

Comparison between the diffraction patterns of samples II and IV shows that the overall picture of the mineralogical composition with the introduction of PFM practically does not change. Compared to a sample that contains only fly ash (diffraction patterns III), when replacing 12.5% of it with blast furnace granulated slag (diffraction patterns IV), there is a slight decrease in the intensity of calcium hydroxide lines and an increase in the amount of calcium hydrosulfoaluminate (ettringite) after 28 days of hardening.

Since with the introduction of fly ash, the rate of $Ca(OH)_2$ formation increases, the increased concentration of calcium hydroxide, which acts as an activator of mineral additives, leads to partial hydrolysis of their glassy component. The interaction of active alumina with calcium hydroxide promotes the formation of ettringite. Its needle crystals, which are formed at an early stage of structure formation, form a crystalline skeleton reinforcing a cement stone, which in turn provides accelerated kinetics of a set of the early strength of CC.

Hydrate formations in the form of thin crystals of calcium hydrosilicate, as well as elongated ettringite crystals, grow in the pores, contributing to their colmatization and increase in stone strength (Figure 1.8). The phase composition of cement stone based on CCs with PFM additives is characterized by an increased content of low-basic hydrosilicates, as well as calcium hydrosulfoaluminate crystals, which form a homogeneous structure with increased density.

The nature of the modification of cement stone with PFM additives is largely determined by a change in the parameters of its pore structure. A universal method for determining the pore structure of cement stone is the method of water absorption based on the phenomenon of capillarity that allows determination of integral (imaginary porosity) and differential (indicators of average size and uniformity of pore size) parameters [42].

To determine the porosity parameters of the cement stone by the kinetics of water absorption, cubic samples were made with a rib size of 70.7 mm. Samples after 28 days of hardening were dried at 105°C–110°C to constant weight. Samples were weighed after 0.25, 1, 6, and 24 hours.

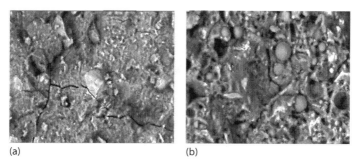

FIGURE 1.8 The microstructure of the stone based on (a) CEM I and (b) CC at 28 days (×1000).

The parameters of the pore structure of cement stone are calculated by a function of the type:

$$W_\tau = W_{max}\left[1 - e^{-\bar{\lambda}\tau^\alpha}\right], \qquad (1.11)$$

where W_{max} is the conditional value of maximum water absorption, $\bar{\lambda}$ is the coefficient characterizing the average size of the capillaries, and α is the coefficient characterizing the uniformity of capillary sizes.

According to the results of water absorption was calculated:

Average density of samples:

$$\rho_o = \frac{m_o \rho_o}{m_{24} - m_{24}^w}. \qquad (1.12)$$

Volumetric water absorption (imaginary porosity):

$$W_o = W_o \rho_0 = \frac{m_{24} - m_o}{m_{24} - m_{24}^w} \cdot 100. \qquad (1.13)$$

Total porosity:

$$P_i = \left(1 - \frac{\rho_b}{\rho}\right) \cdot 100\%. \qquad (1.14)$$

In the above formulas, m_0 and m_{24} are the mass of the sample before immersion in water and 24 hours after immersion is determined in air, respectively, m_{24}^w is the mass of the sample 24 hours after immersion determined in water, ρ is the density of the cement stone, and ρ_b is the bulk density of the cement stone.

The results of determining the water absorption of cement stone are shown in Figure 1.9, and the parameters of the pore structure are provided in Table 1.5.

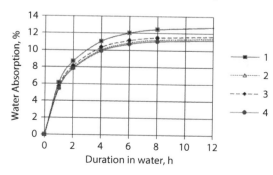

FIGURE 1.9 Kinetics of water absorption of cement-ash stone with PFM additives (ash = 40% by weight of CC): (1) cement-ash paste with Sika VC 225 = 0.7%; S_{sp} = 330 m²/kg; (2) with the addition of PFM$_2$ = 1.5%; S_{sp} = 460 m²/kg; (3) with the addition of PFM$_1$ = 0.7%; S_{sp} = 460 m²/kg; and (4) with the addition PFM$_3$ = 1.7%; S_{sp} = 460 m²/kg.

TABLE 1.5
The Main Parameters of the Pore Structure of Cement-Ash Stone with Additives PFM

Modifier	Amount by Weight CC (%)	Porosity, % Integrated P_i	Porosity, % Open W_0^{max}	Average Pore Size Indicator $\bar{\lambda}$	Pore Size Uniformity Index Indicator α
PFM_1	0.4	13.5	9.9	1.21	0.79
	0.7	14.8	11.1	1.33	0.79
	1.0	15.9	11.1	1.33	0.82
PFM_2	1.0	15.0	10.5	1.23	0.82
	1.5	16.3	10.8	1.26	0.84
	2.0	17.2	11.1	1.25	0.86
PFM_3	1.4	13.9	10.5	1.21	0.79
	1.7	15.0	10.8	1.33	0.81
	2	16.1	11.1	1.33	0.79

Data analysis in Table 1.5 shows that the introduction of PFM additives changes the parameters of the pore structure of cement-ash stone. In all cases, with the introduction of PFM additives, there is pore redistribution toward an increase in closed pores and a decrease in open pores available for saturation with water. At the same time, there is a clear tendency toward a decrease in the average pore size and increase in their uniformity.

Knowing the degree of hydration of the cement stone, it is possible to calculate its relative density, total, capillary, and gel porosity from the calculated dependencies given in Table 1.6.

TABLE 1.6
Estimated Dependencies of Relative Density and Main Type Porosities for Cement-Ash Stone with Additives PFM

Indicator	Estimated Dependence	
Relative density (g/cm³)	$d = \dfrac{1 + 0.23\alpha_{dh} - \rho_c/100}{1 + \rho_c W/C}$	(1.15)
	where α_{dh} is the degree of hydration of cement (%); ρ_c is the true cement density, $\rho_c = 2.9$ g/cm³; W/C is the water-cement ratio	
Total porosity (%)	$P_{tp} = \rho_c \dfrac{(W/C - 0.23\alpha_{dh}/100)}{1 + \rho_c W/C} 100$	(1.16)
Gel porosity (%)	$P_{gp} = \dfrac{0.19\alpha_{dh}\rho_c/100}{1 + \rho_c W/C} 100$	(1.17)
Capillary porosity (%)	$P_{cp} = \rho_c \dfrac{(W/C - 0.42\alpha_{dh}/100)}{1 + \rho_c W/C} 100$	(1.18)

1.4 CONSTRUCTION AND TECHNICAL PROPERTIES OF MODIFIED ASH COMPOSITE CEMENTS AND MORTARS BASED ON THEM

1.4.1 MATERIAL COMPOSITION

CC (to EN 197-1) include cements containing, in addition to blast furnace slag, other mineral additives. The traditional composition for such cements is clinker-gypsum-slag-fly ash. Fly ash as a component of cement should not contain more than 2.5% free CaO, more than 3% alkaline oxides, and the loss on ignition should not exceed 5%. As is known, blast furnace granulated slag as part of an ash-containing composite binder serves not only as an active mineral additive, but can also as a compensator for the negative effects of alkalis, sulfuric anhydride, and free calcium oxide, which are in ash [4]. These components are not only absorbed by slag, but also contribute to an increase in the activity of the latter.

To determine the material composition of finely ground ash binder and the possible boundaries of its variation, trial experiments were performed at the first stage. Binders were made by compatible grinding of Portland cement CEM I, fly ash, blast furnace granulated slag, and modifying additives. The clinker content in CC varied in the range of 50% to 70%, fly ash from 15% to 50%, and blast furnace slag from 10% to 25%. Milling was performed up to a specific surface area of 550 to 600 m^2/kg. For obtained cements, normal consistency, setting time, compressive strength, and bending strength at 2, 7, and 28 days were determined. The results are shown in Table 1.7 and Figure 1.10. Polycarboxylate (Sika VC 225, points 1 to 10, in the amount of 0.7%) and naphthalene formaldehyde (SP-1 points 11 to 19, in the amount of 1.5%) types of SP were used.

An analysis of the data obtained established that the binder that contains blast furnace slag in an amount of 10% to 25% in all cases have higher activating and bending strength compared to a binder that included only cement and fly ash. The optimum content of slag that provides high strength is 10% to 20%. In further studies, a three-component binder was considered, which was prepared by joint grinding of Portland cement of I type containing up to 20% slag, fly ash, and grinding intensifier. During the experiments, the highest compressive and bending strength was obtained using the Sika VC 225—polycarboxylate type SP. The addition of this SP due to its high water-reducing ability leads to a corresponding decrease in the normal consistence of cement and W/C. To compare the effectiveness of SPs of various types, several experiments were conducted with SP-1.

The optimum can be considered the content of polycarboxylate SP in the binder up to 0.7%. So, with an increase in its content from 0.4% to 0.7%, the activity of the binder increases by 25% to 30% in all periods of hardening. A further increase in the amount of additive is impractical, since the strength does not increase significantly. Bending strength also significantly depends on the consumption of SP and increases to 35% with an increase in the consumption of SP from 0.4% to 0.7%. The activity

TABLE 1.7
Properties of Modified Ash Containing CC

No	Binder Composition (%)			Normal Consistency (%)	Bending Strength (MPa, Age, Days)			Compressive Strength (MPa, Age, Days)		
	Cem	Ash	Slag		2	7	28	2	7	28
1	100	—	—	29.00	4.4	6.1	7.9	31.6	44.2	63.2
2	50	50	—	18.25	4	5.2	6.2	24.5	34.2	48.9
3	50	25	25	18.50	4.3	5.9	6.5	26.7	38.4	55.3
4	50	37.5	12.5	18.50	4.2	5.7	6.3	26.1	36.5	52.1
5	60	40	—	17.25	4.3	5.9	7.5	27.1	37.9	54.1
6	60	20	20	17.50	4.5	6.1	7.8	29.2	42.0	58.4
7	60	30	10	17.50	4.3	5.9	7.6	28.6	40.0	57.1
8	70	30	—	17.75	4.7	6.2	7.5	31.6	44.2	63.2
9	70	15	15	18.00	4.8	6.3	8.3	34.7	49.9	69.3
10	70	22.5	7.5	18.00	4.8	6.3	7.9	33.6	47.0	67.1
11	50	50	—	22.00	3.5	4.6	6.6	22.2	31.1	44.5
12	50	25	25	23.00	3.8	4.9	6.7	24.3	34.4	48.5
13	50	37.5	12.5	22.75	3.7	4.8	6.6	23.7	33.2	47.4
14	60	40	—	20.50	3.9	5	6.8	24.1	33.7	48.1
15	60	20	20	21.00	4	5.2	7	26.0	36.4	52.0
16	60	30	10	20.75	4	5.1	6.9	25.4	36.6	50.8
17	70	30	—	21.00	4.1	5.1	6.8	28.4	39.8	56.9
18	70	15	15	21.75	4.3	5.4	6.9	31.2	44.9	62.4
19	70	22.5	7.5	21.50	4.2	5.3	7.2	27.2	42.3	60.4

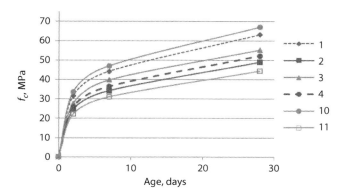

FIGURE 1.10 Compressive strength of cement-ash binders from the content of components (designation according to Table 1.7).

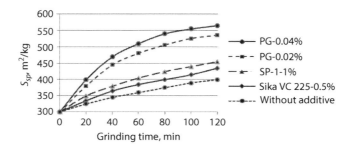

FIGURE 1.11 The effect of additives on the kinetics of grinding CC. The composition of the CC:PC, 50%; ash, 37.5%; slag, 12.5%; and gypsum, 3.5% (SO_3).

and bending strength when using SP-1 in an amount of 1.5% is at this level, as is the binder of modified 0.4% Sika VC 225.

Studies have been conducted on the effects of propylene glycol (PG) additives on the kinetics of grinding CC. For comparison, the effect of SP upon their introduction during the grinding of cement was studied as well. The grinding was carried out in a laboratory ball mill. An analysis of the results showed that the most significant effect on the specific surface of S_{sp} during cement grinding is provided by additives of grinding intensifiers (Figure 1.11). A somewhat lesser effect is observed with the introduction of SP additives, and SP naphthalene-formaldehyde type SP-1 has a more significant effect on the grinding kinetics than SP Sika VC 225. It is explained by their principle of action. As is known, the adsorption effect of surfactant additives (the effect of P.A. Rebinder) depends on the nature of surfactant (SAS) and their ability to interact with cement powder during fine grinding.

As a result of intensification of grinding, PG additives significantly change the grain composition of cement. It is enriched with the finest fractions (up to 10 μm) as shown in Table 1.8.

1.4.2 Properties of Mortar Mixtures

According to the data of [44,45] the ability of a cement gel to hold a certain volume of water is determined by the intensity of the interaction forces between the

TABLE 1.8
Grain Composition of Ash Composite Cements

Additive		Fraction Content (%)				
Type	CC (%)	<10 μm	10–20 μm	20–40 μm	40–60 μm	>60 μm
—	—	25.3	35.5	21.7	13.4	4.1
Sika VC 225	0.5	27.3	37.1	19.6	12.8	3.2
SP-1	1.0	38.7	34.6	14.4	10.6	1.7

particles: the denser they are, the thin shells are packed, the smaller the structural cells between the solvated cement particles, and the more water is held in them. The *water-retention capacity* of cement gel increases with increasing water demand of cement, which can be indirectly determined by normal consistency.

The normal consistency of CCs is influenced by the content of fly ash and SP. The normal consistency of the cements is also affected by their specific surface [46].

To study the influence of these factors on normal consistency CCs and the corresponding quantitative estimates, a three-level three-factor plan B_3 [40] was implemented, the planning conditions of which are given in Table 1.9, and the experimental results are in Table 1.10.

The binder was made by general grinding of an industrial cement together with fly ash with the introduction of one of the components of a PFM—PG. The content of slag in the binder was provided in an amount of 10% to 15%. The influence of two PFMs, PFM_1 and PFM_2, which differ in the type of SP, was ensured. The PFM_1 modifier contained polycarboxylate SP Sika VC 225 and the PFM_2 modifier contained SP-1 naphthalene formaldehyde type SP-1. The content of the intensifier grinding in the composition of the PFM was the same, 0.04% of the amount of binder. Cement paste was prepared in a laboratory mixer and normal consistency was determined according to standard technique.

Statistical processing of experimental data made it possible to obtain mathematical models of the *water demand* of the investigated fine-ground ash-containing CCs in the form of polynomial regression equations:

When using PFM_1:

$$NC_1 = 20.18 + 0.15x_1 + 0.5x_2 - 2.05x_3 - 0.063x_1x_2 \\ + 0.063x_1x_3 + 0.188x_2x_3 + 0.125x_1^2 - 0.125x_2^2 - 0.125x_3^2 \quad (1.19)$$

When using PFM_2:

$$NC_2 = 22.77 - 0.025x_1 + 0.6x_2 - 1.525x_3 \\ + 0.25x_2x_3 + 0.206x_1^2 + 0.169x_2^2 - 0.206x_3^2 \quad (1.20)$$

TABLE 1.9
Experimental Planning Conditions

Factors		Variation Levels			
Natural	Coded	−1	0	+1	Interval
Ash content in CC (ash, %)	x_1	30	40	50	10%
Specific surface (S_{sp}), m²/kg	x_2	350	450	550	100
Content PFM_1/PFM_2 (%)	x_3	0.4/1.0	0.7/1.5	1/2.0	0.3/0.5

TABLE 1.10
Normal Consistence in CC Depending on the Ash and PFM Content and Specific Surface Area

	The Natural Values of the Factors			Normal Consistency of Cement (%)	
No.	Ash (%)	S_{sp} (m²/kg)	PFM[a] (%)	PFM_1	PFM_2
1	50	550	1.0/2.0	19.00	22.00
2	50	550	0.4/1.0	22.5	24.5
3	50	350	1.0/2.0	17.5	20.25
4	50	350	0.4/1.0	22.0	23.75
5	30	550	1.0/2.0	18.75	21.75
6	30	550	0.4/1.0	22.5	24.25
7	30	350	1.0/2.0	17.25	20.00
8	30	350	0.4/1.0	21.50	23.5
9	50	450	0.7/1.5	20.0	22.0
10	30	450	0.7/1.5	20.0	23.25
11	40	550	0.7/1.5	20.5	23.50
12	40	350	0.7/1.5	20.0	22.50
13	40	450	1.0/2.0	18.0	21.00
14	40	450	0.4/1.0	22.0	24.25
15	40	450	0.7/1.5	20.25	22.50
16	40	450	0.7/1.5	20.0	22.50
17	40	450	0.7/1.5	20.25	22.75

Note: Normal consistency of the original cement, 27%.
[a] The numerator is PFM_1 and the denominator is PFM_2.

An analysis of the obtained models indicates that, as expected, the most significant decrease in normal consistency is observed with the introduction of Sika VC 225—polycarboxylate type SP in an amount of 1% by weight of the binder (Figure 1.12). The normal consistency of such cement is reduced to 17% to 18%, that is, a decrease in its water demand is 40% to 42%. In the case of the introduction of SP naphthalene formaldehyde type SP-1 in an amount of 2.0%, it decreases by 18% to 20% (Figure 1.13). A decrease in the consumption of polycarboxylate type SP to 0.7% and naphthalene formaldehyde type to 1.5% does not lead to a significant decrease in the water-reducing ability of SP additives.

As expected, a significant effect on normal consistency cement paste renders an increase in the specific surface of CCs with a content of SP in the studied interval and an increase in S_{sp} from 350 to 550 m²/kg.

It is known that the introduction of fine fly ash helps to reduce the water demand of mortars and concrete. This reduction can be explained by a decrease in the amount of capillary water, which is contained by aggregated particles, and is characterized by a more significant effect compared to an increase in adsorbed film moisture with an increase in the specific surface [46].

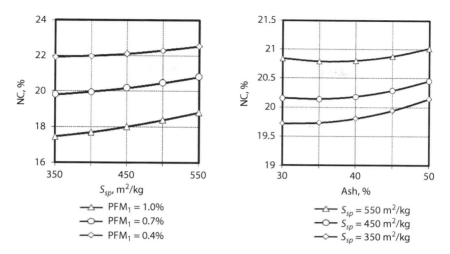

FIGURE 1.12 Dependence of normal consistence of CC on the content of fly ash, PFM-1, and specific surface.

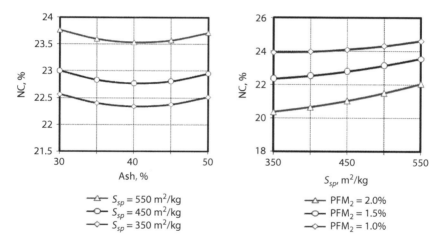

FIGURE 1.13 Dependence of the normal consistency of CC on the ash content, PFM_2, and specific surface.

Figure 1.14 shows the obtained experimental data on the water demand of mortar mixtures with the ratio of sand:binder = 3:1 with additives of modifiers and without them. The water content was adjusted to the cone spreading of the mortar mixture at 135 mm. In this case, the cone is spreading on the shaking table, the mobility of the mixture by immersion of the standard cone, as shown by previous studies, is 8 to 10 cm.

The influence of the amount of ash in the binder on the water demand of mortar mixtures is extreme in nature, which persists with the introduction of modifiers.

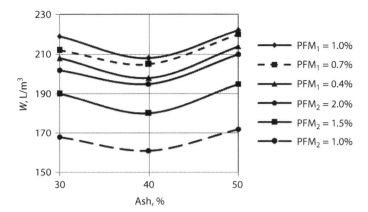

FIGURE 1.14 Dependence of the water demand of mortars based on CC on composition factors ($S_{sp} = 550 \pm 15$ m²/kg).

With the optimum amount of ash to reduce the water demand of the mixtures, a lower modifier content of 0.7% and 1.5% based on the polycarboxylate and naphthalene formaldehyde types of SP, respectively, is required.

The analysis of Figure 1.14 shows that when the content of fly ash in the binder is up to 40%, there is a certain characteristic tendency to decrease the water demand of mortar mixtures, and with a further increase in the consumption of fly ash, water demand increases.

The water demand of the cement mortars is noticeably affected by the sand content and fineness. This position, known in the technology of concretes and mortars [47], is also confirmed for modified mortars based on ash-containing CC.

From Figure 1.15 it follows that with an increase in n (mass ratio of sand and cement), the water demand of mixtures increases.

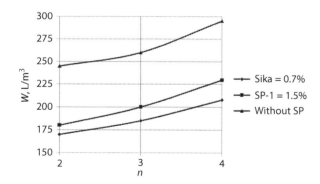

FIGURE 1.15 The influence of the sand-cement ratio (n) of the mortar mixture on its water demand (ash = 40%).

From the data obtained, it can be concluded that the plasticizing effect of SPs is more pronounced in "fatty" mixtures with a large amount of cement-ash binder.

The *water separation* of mortar mixtures is the result of sedimentation compaction and precipitation of solid particles. It is due to the water-retention capacity of the binder and its individual components and depends on the water demand of the aggregates and their content in the mixture [48]. Considering the high dispersion of the binder, it can be expected that filled pastes and mortar mixtures should have a high water-retention capacity.

Coefficient of water separation (K_w, %) of the binder was determined, which is equal to the volume of water, separated:

$$K_w = \frac{a-b}{a} \cdot 100\%, \tag{1.21}$$

where a is the initial volume of the cement slurry in cm³ and b is the volume of cement slurry that has settled in cm³.

Then the water-retention capacity of the suspension:

$$W_{wr}^c = 100 - K_w \tag{1.22}$$

The experiments were performed in duplicate. The dependences of water separation of modified fine-ground cement-ash binders on technological factors are shown in Figures 1.16 and 1.17.

An analysis of the obtained data allows concluding that with the introduction of fly ash, the water-retention capacity of the binder increases with an increase in its content and specific surface. In turn, a significant decrease in water separation is observed with an increase in the specific surface to 550 m²/kg.

With an increase in the specific surface, the positive effect of the fly ash content in the CC composition on water separation increases with an increase in the specific surface of cement to 450 m²/kg and content of PFM based on the polycarboxylate SP.

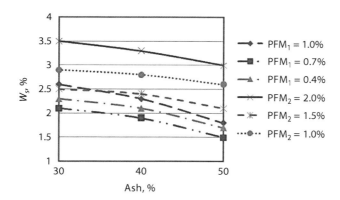

FIGURE 1.16 Dependence of water separation (W_s) of CC on the content of modifier and fly ash (S_{sp} = 465–535 m²/kg).

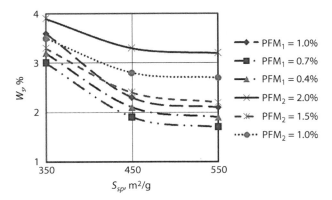

FIGURE 1.17 Dependence of water separation (W_s) CC on the specific surface and the type and content of the modifier (ash = 40%).

A study of cement-ash suspensions showed that during their formation, calcium hydroxide is absorbed on the surface of ash particles and $Ca(OH)_2$ films are formed separated by thin layers of water [4,46]. Hydroxyl ions formed mainly upon dissociation of $Ca(OH)_2$ penetrate on the vitreous phase of the ash, thus causing polarization of the Si-O-Si and Al-O-Si bonds. The Ca^{2+} cations that separated upon dissociation from the solid phase form the outer lining of the double layer facing the liquid phase. According to the theory of the double electric layer, part of the counter ions are in direct contact with the ions of the solid phase, forming a dense layer, the other part is a diffuse layer [24,49]. Electrokinetic phenomena caused by the ions of the diffusion layer are characterized by the space charge of the liquid phase or the ξ potential. With an increase in the excess of ions in the diffusion part, the ξ potential increases, and the transition of ions of the diffuse layer into a dense layer is accompanied by its decrease.

According to modern concepts, the sedimentation stability of disperse systems increases with increasing ξ potential [24]. Saturation of pastes with ash leads to an increase in their ξ potential and an increase in sedimentation stability.

The introduction of SP and rarefaction of pastes is accompanied by significant drop of the ξ potential of pastes, which becomes insensitive to changes in the investigated range of composition.

Air entraining is one of the main factors affecting the strength, uniformity, and durability of concrete and mortar. It is known that each percent of the involved air leads to a decrease in strength by 5% to 10%. The air is drawn in during mixing, unloading, and laying the mixture and depends on the composition, the characteristics of the materials, and the addition of modifiers. The introduction of SPs into the mixtures allows increasing air entraining up to 3% [50]. However, such mixtures quickly lose drawn air due to their low viscosity. The use of fly ash, which is subjected to additional grinding, increases the viscosity of cement systems, which in turn contributes to the retention of entrained air [51]. It is known that polycarboxylate SPs are characterized by increased air retention compared to the naphthalene

TABLE 1.11
The Results of the Experiments on the Determination of Air Entrainment

No.	The Natural Values of the Factors			Air Entrainment of Mortar with PMF (%)	
	Ash (%)	S_{sp} (m²/kg)	PFM (%)	PFM$_1$	PFM$_2$
1	50	550	1.0/2.0	7.4	3.9
2	50	550	0.4/1.0	5.8	3.3
3	50	350	1.0/2.0	4.7	2.8
4	50	350	0.4/1.0	3.8	2.1
5	30	550	1.0/2.0	7.0	3.7
6	30	550	0.4/1.0	5.3	2.9
7	30	350	1.0/2.0	4.3	2.5
8	30	350	0.4/1.0	2.5	1.5
9	50	450	0.7/1.5	4.7	2.8
10	30	450	0.7/1.5	4.4	2.4
11	40	550	0.7/1.5	6.6	3.5
12	40	350	0.7/1.5	3.3	2.1
13	40	450	1.0/2.0	5.6	3.0
14	40	450	0.4/1.0	4.1	2.3
15	40	450	0.7/1.5	4.9	2.6
16	40	450	0.7/1.5	4.9	2.7
17	40	450	0.7/1.5	4.8	2.7

formaldehyde type SP, which leads to a decrease in the "distance factor" in concretes and mortars [52]. At the same time, the effect of polycarboxylate type additives on air entrainment has not been adequately studied and requires further in-depth research.

Air entrainment in mortars based on CCs with PFM additives was determined by volumetric method. The planning conditions, matrix and the results of experiments to determine air entrainment are given in Tables 1.9 and 1.11.

Experimental-statistical models of air entrainment mortars with PFM additives:

$$AE_1 = 4.77 + 0.29x_1 + 1.35x_2 + 0.75x_3 - 0.1x_1x_2 \\ - 0.125x_1x_3 + 0.075x_2x_3 + 0.120x_1^2 - 0.27x_2^2 + 0.17x_3^2;\quad(1.23)$$

$$AE_2 = 2.65 + 0.19x_1 + 0.63x_2 + 0.38x_3 - 0.038x_1x_2 \\ - 0.063x_1x_3 - 0.038x_2x_3 + 0.021x - 0.179x_2^2 + 0.029x_3^2.\quad(1.24)$$

The integral effect of air entrainment of ash-containing CC along with the content of modifier additives is significantly affected by the fineness of cement grinding. With an increase in S_{sp}, the negative effect of the fly ash content on the intensity of air entrainment in mortar mixtures is weakened. This is evidenced by significant effects of the interaction of factors in models (1.23 and 1.24).

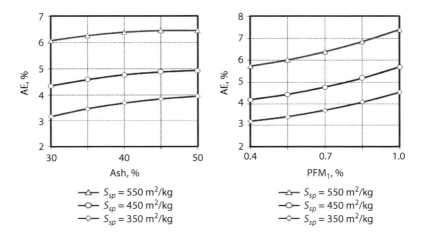

FIGURE 1.18 Dependence of air entrainment CC on the content of ash, PFM_1, and ash specific surface.

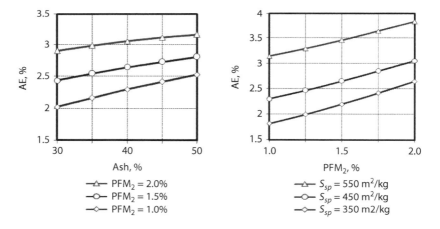

FIGURE 1.19 Dependence of air entrainment CC on the content of ash, PFM_2, and ash specific surface.

Analyzing the obtained data, it can be concluded that an increase in ash consumption in the studied intervals leads to a decrease in the amount of air involved in mortars ash binders. An increase in the dispersion of the binder, on the contrary, contributes to an increase in the amount of air involved. So, with an increase in ash content from 30% to 50% (Figures 1.18 and 1.19), air entrainment is reduced by 0.5% in mixtures.

Thus, the necessary air entrainment, as well as other properties of mortars based on modified fine-ground ash binder, can be achieved by controlling the ash consumption and dispersion of cement.

Composite Ash-Containing Cements and Effective Concrete Based on Them   33

1.4.3 STRENGTH INDICES OF ASH-CONTAINING COMPOSITE CEMENTS

The main studies of strength characteristics in order to reduce the number of experiments were performed using mathematical planning of the experiment. Accordingly, for this, a three-level three-factor plan B_3 was used, and the planning conditions are given in Table 1.9 [40].

Along with compressive and tensile strengths, an important characteristic is the criterion for crack resistance of mortars, just as for concrete. Crack resistance can be expressed as the ratio of bending strength to compressive strength ($f_{m,tb}/f_m$).

In the course of studies, at each point in the plan, a cement-sand mortar with a modified cement-ash binder composition was prepared: at n = binder:sand—1:3. W/C was determined to achieve a spread of mortar cone on the shaking table of at least 135 mm, the strength of the beam—samples on compression and bending determined at 2, 7, and 28 days. The planning conditions and the experimental results obtained using PFM_1 and PFM_2 are presented in Tables 1.9 and 1.12.

After processing and statistical analysis of the experimental data, mathematical models of the strength of the mortars were obtained based on the studied modified binders under compression (f_m) and bending ($f_{m,tb}$) in the form of polynomial regression equations.

The coefficients of experimental-statistical models of the strength of ash-containing CC are given in Tables 1.13 and 1.14.

Graphic dependencies illustrating the influence of technological factors on the strength of ash-containing CC on compression and bending are shown in Figures 1.20 through 1.23.

As follows from Table 1.12, the introduction of PFM solutions and fly ash in the mortars positively affects the criteria for crack resistance. An increase in crack resistance value is characteristic for modified mortars at all hardening periods.

Analysis of the graphs in Figure 1.20 shows that the specific surface has a significant effect on the early strength of finely ground cement-ash binders. With an increase in S_{sp}, the strength increases by 30% to 45%. The increased dispersion of CC is better manifested when the consumption of fly ash, 30% compared to 50%. The optimal consumption of Sika VC 225 SP in the modifier is 0.7% to 1%, and SP-1 is 1.5%, which leads to an increase in early strength by 25% to 35%. A further increase in the amount SP of the polycarboxylate type additive is impractical, since the strength does not increase significantly, and the SP-1 SP, as is known, inhibits hardening at an increased dosage. An increase in the content of Portland cement in CC from 50% to 70% leads to an increase in strength by 20% to 30%.

From Figure 1.21, it follows that the content of fly ash in the binder has the greatest effect on the strength of CC, with an increase as the strength of CC decreases. Under such conditions, to increase the activity of the CC, it is advisable to introduce an SP, which reduces its water demand. With an increase in the consumption of SP, W/C decreases and the strength increases accordingly.

The optimal consumption of SP in the binder when using PFM_1 is 0.7%, which leads to an increase in strength by 20% to 25% compared with the consumption of PFM_1 at 0.4%. In the case of dosing PFM_2, it is optimal to use an additive at 1.5%, which leads to an increase in activity by 15% to 20% compared with a consumption

TABLE 1.12
Planning Matrix and Experimental Values of the Strength Characteristics of CC with PFM Additives

Factors				Strength at Days, MPa						$f_{m,tb}$
				Bending			Compressive			f_m
Ash (%)	S_{sp} (m²/kg)	PFM (%)	W/C	2	7	28	2	7	28	
50	550	1.0	0.28	5.9	8.0	9.5	22.1	42.2	51.8	0.18
		2.0	0.31	4.7	6.8	8.1	17.7	38.0	46.6	0.17
50	550	0.4	0.33	4.8	5.7	8.3	19.2	32.8	39.7	0.21
		1.0	0.36	4.3	4.8	7.1	17.3	29.5	35.7	0.20
50	350	1.0	0.24	4.5	6.0	7.9	16.2	36.2	49.8	0.16
		2.0	0.26	3.6	5.1	6.7	13.0	32.6	44.8	0.15
50	350	0.4	0.24	4.5	6.0	7.9	16.2	36.2	49.8	0.16
		1.0	0.26	4.1	5.1	6.7	14.6	32.5	44.7	0.15
30	550	1.0	0.29	6.7	8.8	10.3	29.8	44.2	55.1	0.19
		2.0	0.32	5.4	7.5	8.8	23.8	39.8	49.6	0.18
30	550	0.4	0.34	5.4	5.9	8.6	22.4	33.8	41.2	0.21
		1.0	0.37	4.9	5.0	7.3	20.2	30.4	37.1	0.20
30	350	1.0	0.26	4.8	6.8	8.2	18.9	40.5	53.1	0.15
		2.0	0.29	3.8	5.8	7.0	15.1	36.5	47.8	0.15
30	350	0.4	0.31	4.4	6.6	7.6	16.5	30.1	39.9	0.19
		1.0	0.34	4.0	5.6	6.5	14.9	27.1	35.9	0.18
50	450	0.7	0.26	6.0	7.4	8.8	17.3	37.8	50.3	0.17
		1.5	0.29	5.4	6.3	7.5	15.6	34.0	45.3	0.17
30	450	0.7	0.28	6.4	8.3	9.5	20.8	42.0	60.2	0.16
		1.5	0.31	5.8	7.1	8.1	18.7	37.8	54.2	0.15
40	550	0.7	0.30	6.4	8.1	10.0	21.7	38.2	51.0	0.20
		1.5	0.33	5.8	6.9	8.5	19.5	34.4	45.9	0.19
40	350	0.7	0.27	4.2	6.7	8.1	19.4	38.0	50.4	0.16
		1.5	0.30	3.8	5.7	6.9	17.5	34.2	45.4	0.15
40	450	1.0	0.23	6.0	7.6	8.7	19.5	38.7	52.6	0.17
		2.0	0.25	4.8	6.5	7.4	15.6	34.8	47.3	0.16
40	450	0.7	0.27	6.3	7.7	9.3	20.1	34.9	58.4	0.16
		1.5	0.30	5.7	6.5	7.9	18.1	31.4	52.6	0.15
40	450	0.7	0.28	6.2	7.5	9.1	18.9	34.5	57.8	0.16
		1.5	0.30	5.6	6.4	7.7	17.0	31.1	52.0	0.15
40	450	0.7	0.27	6.3	7.8	9.4	19.8	35.4	59.0	0.16
		1.5	0.30	5.7	6.6	8.0	17.8	31.9	53.1	0.15

Note: The numerator is PFM_1 and the denominator is PFM_2.

of 1%. However, a further increase in the amount of additive is impractical, since with a decrease in W/C, the strength of mortars does not increase significantly. An increase in the specific surface to 450 m²/kg leads to an increase in activity by 25% to 35% compared to 350 m²/kg; however, with a specific surface of about 550 m²/kg, a significant increase in activity is not observed.

TABLE 1.13
The Coefficients of Experimental-Statistical Models of the Strength of Mortars Based on Ash Containing CC with the Addition of PFM$_1$

Strength	Age, Days	b_0	b_1	b_2	b_3	b_{12}	b_{13}	b_{23}	b_1^2	b_2^2	b_3^2
f_m	2	19.1	−1.74	2.8	1.7	−0.99	−0.86	0.99	0.39	1.89	−1.32
	7	36.31	−0.54	1.02	3.84	−0.6	−1.43	1.18	2.68	0.88	−2.62
	28	56.14	−0.81	−0.42	4.87	−1.43	−1.88	1.6	0.97	−3.58	−6.47
$f_{m, tb}$	2	6.22	−0.2	0.68	0.3	−0.15	−0.08	0.25	0.03	−0.69	−0.27
	7	7.79	−0.33	0.44	0.6	0.5	−0.1	0.63	−0.02	−0.47	−0.57
	28	9.19	−0.18	0.7	0.4	−0.14	−0.14	0.29	0.04	−0.06	−0.66

TABLE 1.14
Coefficients of Experimental-Statistical Models of Strength of Mortars Based on Ash-Containing CC with the Addition of PFM$_2$

Strength	Age, Days	b_0	b_1	b_2	b_3	b_{12}	b_{13}	b_{23}	b_1^2	b_2^2	b_3^2
f_m	2	17.22	−1.46	2.35	0.47	−0.83	−0.65	0.68	0.3	1.65	−2.63
	7	32.68	−0.49	0.92	3.46	−0.54	−1.28	1.06	2.41	0.79	−2.36
	28	50.53	−0.73	−0.38	4.38	−1.28	−1.69	1.44	0.88	−3.22	−5.77
$f_{m, tb}$	2	5.59	−0.17	0.58	−0.01	−0.13	−0.05	0.18	0.04	−0.77	−0.53
	7	6.63	−0.28	0.37	0.51	0.04	−0.09	0.53	−0.01	−0.4	0.48
	28	7.81	−0.15	0.6	0.34	−0.12	−0.12	0.24	0.04	−0.05	−0.56

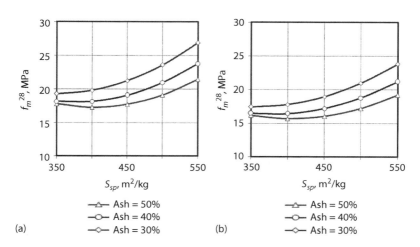

FIGURE 1.20 The influence of technological factors on the strength of CC-based mortars under compression at 2 days with additives (a) PFM$_1$ and (b) PFM$_2$.

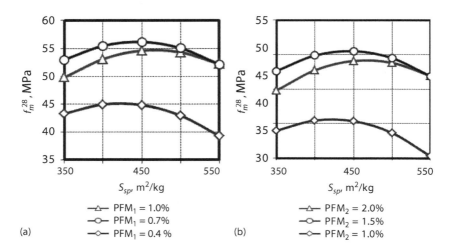

FIGURE 1.21 The influence of technological factors on the strength of CC-based mortars under compression at 28 days with additives (a) PFM_1 (b) та PFM_2.

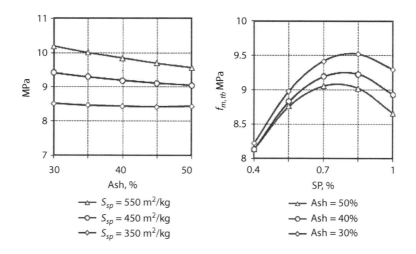

FIGURE 1.22 The influence of technological factors on the bending strength of the mortars at 28 days when using PFM_1.

The main factor that significantly affects the strength of CC during bending is the specific surface, with an increase from 350 to 550 m²/kg, the bending strength increases by 15% to 25%. The increase in the content of fly ash in the CC from 30% to 50% leads to a slight decrease in strength to 10% (Figures 1.22 and 1.23).

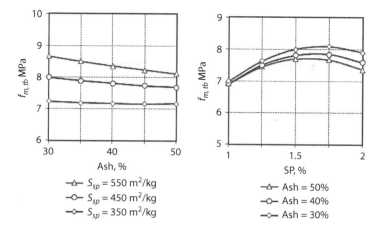

FIGURE 1.23 Influence of technological factors on the bending strength of the mortars at 28 days when using PFM$_2$.

1.4.4 THE ADHESIVE PROPERTIES OF MORTARS BASED ON ASH CONTAINING CC

Increasing adhesion is one of the main tasks in obtaining modified mortars. An analysis of the mechanisms of contact interactions in filled cement systems makes it possible to determine the ways of activation of fillers to increase their adhesion to the binder and improve structure formation.

The adhesion of cement mortars is a complex dependence on many factors, including their strength [4,50]. An increase in the strength of the mortars usually is achieved by increasing the activity of cement and C/W, wherein an increase in C/W is provided by increasing the consumption of cement or by introducing SP, which is not always rational.

Several methods are known for improving the adhesion ability of cement stone with its limited amount in the mortar. One of them is based on a concept that considers cement stone as micro-concrete. According to this concept, it is advisable to increase the dispersion of cement, ensuring its maximum hydration. Grains of cement larger than 40 microns, which are practically not hydrated, are rationally replaced by fillers. Mechanical processes during the grinding of mineral fillers lead not only to an increase in their surface energy, which leads to their increased adhesive activity, but also to an increase in their chemical activity, which also contributes to high adhesive strength upon their contact with a binder. However, it is necessary to consider the property of ground fine powders to rapid deactivation in air due to their high adsorption capacity.

All contact systems are divided into equilibrium and no-equilibrium. In equilibrium systems, the chemical potentials of the components of the contacting phases are the same, the adhesion bonds are localized at the atoms at the surface of each phase, and the free valences of the atoms of bodies located at the boundary are interconnected. In no-equilibrium systems, a chemical reaction occurs at the interface, with a partial break of such bonds in the volume of each phase, intermediates of different types are formed, diffusion takes place, and dissolution of one phase into another. The initial boundary between the phases is broken and a new one is often formed, for example, between one of the phases and the intermediate layer of the chemical compound. When selecting fillers and determining their activation pathways, it is desirable to strive for the formation of chemically no-equilibrium systems with high adhesive strength [4,50].

To create sufficiently strong adhesive bonds in the cement-filler system, it is necessary that the surface energy of the filler was significantly higher than that of cement. This statement is based on the thermodynamic concept of adhesion [47]. According to this concept, the main role in the formation of adhesive strength is assigned to the ratio of the values of the surface adhesion energy of adhesive W_{ad} and substrate W_{sub}. Mandatory is the condition $W_{ad} < W_{sub}$.

When choosing surfactants (SAS) it is necessary to consider the chemical nature of the filler and the binder. One of the necessary conditions for the effectiveness of SAS is their chemisorption interaction ability with the surface of the filler particles. For mineral fillers, which are acidic in nature, the cationic types of SAS are most effective, and for fillers with basic nature, anionic SAS are the most effective.

The influence of the binder composition, as well as the type and content of the studied modifier additives, was studied using mathematical planning of experiments. Planning conditions and research results are given in Tables 1.9 and 1.15.

TABLE 1.15
The Experimental Results in the Study of the Adhesion Strength of Mixtures with Additives PFM$_1$ and PFM$_2$

No.	Factors			Adhesion to Base (MPa)	
	Ash (%)	S_{sp} (m²/kg)	PFM$_1$/PFM$_2$ (%)	PFM$_1$	PFM$_2$
1	50	550	1.0/2.0	1.9	1.7
2	50	550	0.4/1.0	1.5	1.3
3	50	350	1.0/2.0	0.7	0.6
4	50	350	0.4/1.0	0.5	0.4
5	30	550	1.0/2.0	2.2	2.0
6	30	550	0.4/1.0	1.7	1.5
7	30	350	1.0/2.0	0.8	0.7
8	30	350	0.4/1.0	0.7	0.6
9	50	450	0.7/1.5	1.3	1.2
10	30	450	0.7/1.5	1.8	1.6
11	40	550	0.7/1.5	2.1	1.9
12	40	350	0.7/1.5	0.9	0.8
13	40	450	1.0/2.0	1.4	1.2
14	40	450	0.4/1.0	0.9	0.8
15	40	450	0.7/1.5	1.6	1.4
16	40	450	0.7/1.5	1.6	1.4
17	40	450	0.7/1.5	1.7	1.5

The water demand of the mortar mixture was selected to achieve a mobility of 10 to 12 cm. As a result of statistical processing, experimental-statistical models of the adhesive strength of mortars were obtained (1.25 and 1.26).

Experimental-statistical models of adhesive strength:

When using PFM_1:

$$f_{m,adh(1)} = 1.565 - 0.13x_1 + 0.58x_2 + 0.17x_3 \\ - 0.025x_1x_2 + 0.075x_2x_3 + 0.01x_1^2 - 0.39x_2^2 - 0.359x_3^2. \quad (1.25)$$

When using PFM_2:

$$f_{m,adh(2)} = 1.387 - 0.12x_1 + 0.53x_2 + 0.16x_3 \\ - 0.025x_1x_2 + 0.075x_2x_3 + 0.051x_1^2 - 0.32x_2^2 - 0.349x_3^2. \quad (1.26)$$

Based on the obtained models, graphical dependencies are constructed (Figures 1.24 and 1.25).

Based on the effect on the adhesion strength value, the studied factors can be placed in the following order S_{sp} > SP > Ash. An increase in the specific surface from 350 to 550 m²/kg allows an increase in the adhesive strength of the mortar by two to three times or 1, 0, to 1.2 MPa (modifier based on Sika VC 225, Figure 1.24). When using SP-1 (Figure 1.25), the increase in adhesion is smaller, but also significant and is about 0.7 to 1.0 MPa. The optimal dosage of Sika VC 225 is in the range of 0.6% to 0.8% and 1.25% to 1.75% for SP-1. An increase in the quantity of fly ash in cement up to 50% slightly reduces the adhesion strength by 0.2 to 0.3 MPa,

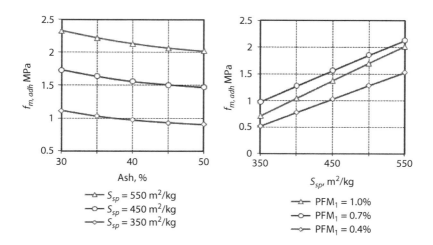

FIGURE 1.24 The dependence of the adhesive strength of mortars on the ash content and specific surface area (PFM-1).

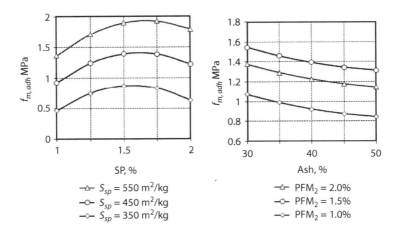

FIGURE 1.25 The dependence of the adhesive strength of mortars on the content of SP and ash (PFM$_2$).

regardless of PFM. The maximum adhesion strength to the base of fine-ground modified cement-ash binders is 2.2 and 2.0 MPa, respectively, when PFM$_1$ and PFM$_2$ are introduced at optimal values of fly ash, SP additive consumption, and specific surface area. An analysis of the obtained data indicates that when applied in mortars of ash-containing CC with PFM additives, including SPs of both naphthalene-formaldehyde and polycarboxylate types and grinding intensifier, it is possible to provide sufficiently high values of adhesive strength in the range of optimal compositions.

1.5 HIGH-TECH CONCRETES BASED ON ASH-CONTAINING COMPOSITE CEMENT

1.5.1 Properties of Concrete Mixtures

Water demand and workability of concrete mixtures: To obtain high-tech concretes of the type high performance concrete (HPC) [54–56], it is necessary to reduce the water-cement ratio to W/C < 0.4, which requires the use of special technological measures. Such measures include the use of SPs and complex additives based on them. For concrete without additives of SPs, at low W/C, water demand constancy rule of the concrete mixture is violated, that is, with an increase in the consumption of Portland cement more than 400 kg/m³ (for fine-ground cement more than 350 kg/m³), it is necessary to simultaneously increase the water consumption of concrete mixes for compensation increases the viscosity of the cement paste. Accordingly, the consumption of cement increases, worsens the structure of concrete, and causes increased shrinkage and destructive phenomena. All this reduces the effectiveness of the use of cement. Therefore, an increase in cement consumption in concrete of more than 550 to 600 kg/m³ is impractical from economic considerations and from the standpoint of improving the technical properties of concrete.

TABLE 1.16
Experiment Planning Conditions

Factors of Influence		Levels Varying			
Natural	Coded	−1	0	+1	Interval
Content PFM_1 (in % of CC by weight)	x_1	0.4	0.7	1.0	0.3
Specific surface area CC, S_{sp} (m²/kg)	x_2	350	450	550	100
Cement consumption C (kg/m³)	x_3	300	400	500	100
Slump cone of concrete mix, S (cm)	x_4	2	13	24	11

Taking into account the results of studies on the effectiveness of PFM of various compositions on the properties of ash CC for research in concrete, PFM_1 was chosen, which contains, along with the intensifier of grinding cement-propylene glykol (PG), a polycarboxylate type SP Sika VC 225.

To study the water demand and workability of concrete mixtures based on composite ash cement with PFM additives, a three-level four-factor plan B_4 was implemented [40], the conditions of which are given in Table 1.16.

The studies consisted of determining the water demand of concrete mixtures with a given workability, various parameters, of the composite binder composition, and its consumption. As initial materials used:

- As components of CC: Portland cement Cem II; fly ash, grinding intensifier—PG;
- Polycarboxylate type SP—Sika VC 225;
- Coarse aggregate—crushed granite with D_{max} = 20 mm;
- Fine aggregate—quartz sand (P_1) with M_f = 1.95.

The composition of the binder was as follows: clinker, 50%; blast furnace slag, 12%; fly ash, 38%; and PG, 0.04%.

The obtained experimental values of water demand are given in Table 1.17.

After processing and statistical analysis of the experimental data, a mathematical model of the water demand of the concrete mixture using ash-containing CC was obtained:

$$W_{c.m.} = 142 - 24.92x_1 + 1.14x_2 + 3.92x_3 + 5.49x_4$$
$$- 0.63x_1x_2 + 0.63x_1x_3 + 1.25x_2x_4 + 0.38x_3x_4 \quad (1.27)$$
$$- 0.68x_1^2 + 1.82x_2^2 + 0.82x_3^2 - 1.18x_4^2.$$

The mathematical model (1.27) is characterized by a significant interaction of the effects of the influence of the specific surface of cement and the consumption of CC, as well as the workability of the concrete mixture. Specifically with an increase in S_{sp}, the water demand of the concrete mix decreases while increasing slump cone

TABLE 1.17
The Results of Experiments to Determine Water Demand of Concrete Mix

No	PFM$_1$ (%)	S_{sp} (m²/kg)	C (kg/m³)	S (cm)	$W_{c.m}$ (L/m³)
		Influence Factors			
1	1.0	550	500	24	134
2	1.0	550	500	2	123
3	1.0	550	300	24	122
4	1.0	550	300	2	112
5	1.0	350	500	24	125
6	1.0	350	500	2	113
7	1.0	350	300	24	118
8	1.0	350	300	2	107
9	0.4	550	500	24	184
10	0.4	550	500	2	172
11	0.4	550	300	24	174
12	0.4	550	300	2	164
13	0.4	350	500	24	172
14	0.4	350	500	2	160
15	0.4	350	300	24	167
16	0.4	350	300	2	157
17	1.0	450	400	13	117
18	0.4	450	400	13	166
19	0.7	550	400	13	148
20	0.7	350	400	13	140
21	0.7	450	500	13	147
22	0.7	450	300	13	139
23	0.7	450	400	24	146
24	0.7	450	400	2	136

and reducing the consumption of cement. A significant non linear effect is observed when analyzing the influence of the specific surface and workability of mixtures. The ranking of quantitative effects of the influence of the studied factors on water demand makes it possible to arrange them according to the decreasing force of the influence $x_1 > x_2 > x_3 > x_4$.

Graphical dependences of the water demand of the concrete mixes on the factors studied are shown in Figure 1.26.

From Figure 1.26, it follows that the PFM-1 content has the greatest influence on the water demand of concrete mixtures, an increase in its consumption from 0.4% to 1% (by weight of the CC) leads to an almost linear decrease in the water demand of the concrete mixture from 170 to 180 L/m³ to 110 to 120 L/m³. The influence of other factors is less significant. With an increase in the workability of the concrete mixture from stiff to cast consistency, the consumption increases by 10 to 15 L/m³. The increased dispersion of the binder (450 to 550 m²/kg) leads to an increase in water consumption of 7 to 10 L/m³ in comparison with the specific surface of

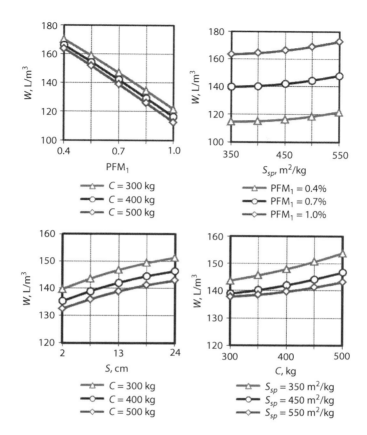

FIGURE 1.26 Influence of composition factors on the water demand of concrete mixtures based on ash-containing CC.

350 to 450 m²/kg. With an increase in dispersion from 350 to 450 m²/kg, the increase in water consumption is smaller compared to S_{sp} at 450 to 550 m²/kg. The CC content is the least significant influence factor on water demand in the studied range. With an increase in its content from 300 to 500 kg/m³, water demand increases by 3 to 6 L/m³ and with increased dispersion increases up to 10 L/m³.

Thus, the conducted studies indicate that to produce cast and self-compacted concrete mixes based on CC in the entire range of cement consumption and its dispersion, a necessary condition is an increased consumption of PFM.

The analysis of the obtained graphs indicates that $S > 22$ cm is achieved with a consumption of SP of 0.7% to 1% and S_{sp} at 350 to 450 m²/kg. The consumption of SP affects the value of the specific surface of cement and the increase leads to a decrease in the slump cone of the concrete mix. Consumption of cement, in turn, does not have a significant effect on slump cone. Determination of water demand and its regulation from changes in the main technological factors can be carried out using a nomogram (Figure 1.27), which is built based on a mathematical model (1.27).

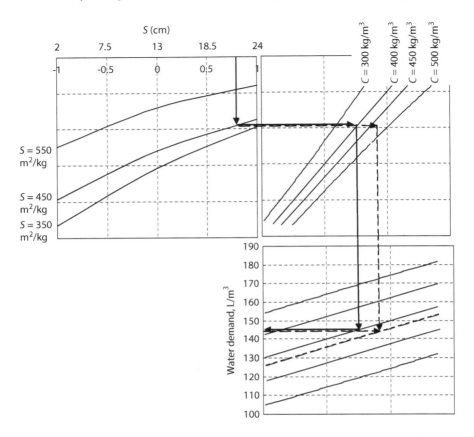

FIGURE 1.27 Nomogram of determination of water demand of concrete mixtures based on ash-containing CC.

As is known, the water demand of concrete mixtures is correlated with the normal consistency of cement. The corresponding experimental data on the dependence of the water demand of concrete mixtures based on CCs with the addition of PFM_1 for various workabilities are given in Table 1.18.

Processing the data given in Table 1.18 gives the graphical dependencies (Figure 1.28) and the regression equation of water demand (Table 1.19).

1.5.2 THE STRENGTH OF CONCRETE

One of the main features of the so-called high-tech concretes is its high strength characteristic. In addition to high compressive strength, high-tech concretes are characterized by increased crack resistance, water impermeability, and frost resistance. These properties also are determined significantly by the strength characteristics of the concrete [57].

The main studies of the effect of CC dispersion, the content of PFM, and W/C on the strength characteristics of concrete based on these cements were performed

Composite Ash-Containing Cements and Effective Concrete Based on Them 45

TABLE 1.18
Dependences of Water Demand of Concrete Mixtures on Normal Consistency (NC) Cement and Maximum Aggregate Size (D_{max})

	Water Requirement of Concrete Mix (kg/m³) at Its Consistency			
	S = 1 to 4 cm		S > 22 cm	
NC of CC, (%)	D_{max} = 20 mm	D_{max} = 5 mm	D_{max} = 20 mm	D_{max} = 5 mm
16	105	135	145	230
20	125	165	175	250
24	150	220	210	275

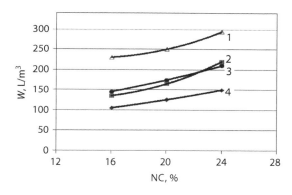

FIGURE 1.28 The influence normal consistence of cement and maximum aggregate size on the water demand of concrete mixtures based on ash-containing CC:

1: D_{max} = 20 mm, S = 1–4 cm; 2: D_{max} = 20 mm, S > 22 cm;

3: D_{max} = 5 mm, S > 22 cm; and 4: D_{max} = 5 mm, S = 1…4 cm.

TABLE 1.19
The Regression Equations for the Water Demand of Concrete Mixtures from Normal Consistency Cement and the Maximum Size of Aggregates

Parameters

D_{max}, mm	Slump Cone	Equation
5	S > 22 cm	$W_1 = 0.781NC^2 - 23.12NC + 400$
5	S = 1…4 cm	$W_2 = 0.781\ NC^2 - 20.62\ NC + 265$
20	S > 22 cm	$W_3 = 0.156\ NC^2 + 1.875\ NC + 75$
20	S = 1…4 cm	$W_4 = 0.156\ NC^2 - 0.625\ NC + 75$

TABLE 1.20
Experiment Planning Conditions

Factors		Variation Levels			
Natural	Coded	−1	0	+1	Interval
W/C	x_1	0.25	0.3	0.35	0.05
Content PFM_1 at CC (%)	x_2	0.4	0.7	1	0.3
S_{sp} (m²/kg)	x_3	350	450	550	100

using mathematical planning of experiments. For this, a three-level three-factor plan B_3 was implemented, the planning conditions of which are given in Table 1.20. In the course of research at each point of the plan, studies have been performed to assess the influence of factors on the strength of concrete.

Standard cubes (10 × 10 × 10 cm) were made and hardened under normal conditions. The compressive strength of specimen cubes was determined at 1, 7, and 28 day and splitting tensile strength at 28 days. The obtained experimental results are presented in Table 1.21.

TABLE 1.21
Experimental Values of the Strength Characteristics of Concrete Based on Ash-Containing CC

No	Factors			Slump Cone	Compressive Strength, in Age, Days, f_c (MPa)			Splitting Tensile Strength $f_{c.sp}$	The Ratio $\dfrac{f_{c.sp}}{f_c}$
	W/C	PFM_1 (%)	S_{sp} (m²/kg)	S (cm)	1 Days	7 Days	28 Days	28 Days	28 Days
1	0.35	1	350	28	21.2	39.8	58.4	4.11	0.07
2	0.35	1	550	27	35.6	39.3	74.2	4.82	0.065
3	0.35	0.4	350	7	22.0	46.2	59.2	4.15	0.07
4	0.35	0.4	550	9	34.7	46.7	74.6	4.84	0.065
5	0.25	1	350	22	25.0	46.6	67.3	4.52	0.067
6	0.25	1	550	24	41.5	64.7	90.0	5.49	0.061
7	0.25	0.4	350	0	26.4	51.9	67.6	4.53	0.067
8	0.25	0.4	550	2	42.9	64.6	91.0	5.53	0.061
9	0.35	0.7	450	20	36.7	53.5	75.1	4.86	0.065
10	0.25	0.7	450	16	37.4	61.6	86.4	5.34	0.062
11	0.30	1	450	24	34.2	58.8	75.5	4.88	0.065
12	0.30	0.4	450	4	29.9	58.4	75.9	4.90	0.065
13	0.30	0.7	350	16	34.5	51.8	67.0	4.51	0.067
14	0.30	0.7	550	19	37.4	64.3	85.3	5.30	0.062
15	0.30	0.7	450	18	37.2	60.1	80.2	5.08	0.063
16	0.30	0.7	450	17	36.3	59.3	79.3	5.04	0.064
17	0.30	0.7	450	17	35.6	58.1	80.1	5.08	0.063

After processing and statistical analysis of the experimental data, mathematical strength models in the form of polynomial regression equations are obtained:

Concrete compressive strength at 1 day (MPa):

$$f_c^1 = 36.72 - 2.31x_1 + 1.68x_2 + 6.3x_3 + 0.36x_1x_2 \\ + 0.72x_1x_3 - 0.21x_2x_3 + 0.21x_1^2 - 4.8x_2^2 - 0.91x_3^2.$$

(1.28)

Concrete compressive strength at 28 days (MPa):

$$f_c^{28} = 8061 - 6.09x_1 + 0.3x_2 + 0.57x_3 + 0.01x_1x_2 \\ + 0.86x_1x_3 + 0.38x_2x_3 + 2.5x_1^2 - 4.28x_2^2 - 3.78x_3^2.$$

(1.29)

The splitting tensile strength of concrete at 28 days (MPa):

$$f_{c,t}^{28} = 5.08 - 0.262x_1 + 0.013x_2 + 0.415x_3 + 0.069x_1x_3 \\ + 0.001x_2x_3 + 0.027x_1^2 + 0.185x_2^2 - 0.172x_3^2.$$

(1.30)

When analyzing mathematical models (1.28 through 1.30), there is a significant interaction of W/C and S_{sp} factors on early and 28-days strengths, while other interactions have a negligible effect. A significant non linear effect of the influence of all factors can be traced at a later age. The ranking of the quantitative effects of the influence of the studied factors on the strength of the compressive concrete indicators during compression allows them to be arranged in this order $x_3 > x_1 > x_2$. The influence of factors on splitting tensile strength is of a similar nature. The determining factor is the specific surface area of the CC.

Graphic dependences of concrete compressive strength at 1 and 28 days, built based on the obtained models, are shown in Figure 1.29, and splitting tensile at 28 days are shown in Figure 1.30.

An analysis of the obtained data (Table 1.21, Figure 1.29) indicates that the compressive strength of the studied concrete at 1 day lies in the range of 20 to 43 MPa and at 28 days in the range of 59 to 92 MPa at a binder consumption of 500 kg/m³ (cement clinker content: 250 kg/m³ concrete mix). The main factor that affects the strength of concrete is the specific surface area of the CC. An increase in S_{sp} from 350 to 550 m²/kg leads to a regular increase in strength of 35% to 50% at all hardening periods. However, at 1 day, concretes on finer dispersed cement (S_{sp} = 450 to 550 m²/kg) have a high strength of 40% to 50% compared to S_{sp} = 350 m²/kg. At a later age, the effect of dispersion is smoothed out. W/C also has a significant effect on early and 28-day strength. An increase in W/C from 0.25 to 0.35 accompanies a decrease in strength by 10 to 20 MPa.

According to the data (Table 1.21) and graphical dependencies (Figure 1.30), the splitting tensile strength of the investigated concrete at 28 days is 4.11 to 5.49 MPa, while the influence of factors in their significance is similar to the effect on compressive strength.

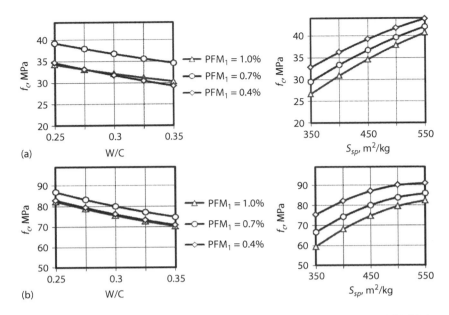

FIGURE 1.29 Dependences of concrete compressive strength in (a) 1 day and (b) 28 days.

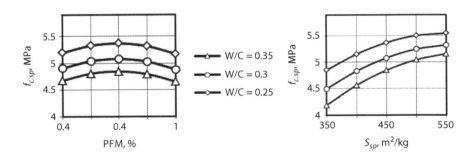

FIGURE 1.30 Dependences of concrete splitting tensile strength at 28 days.

Thus, analyzing the graphical dependencies (Figures 1.29 and 1.30), a region can be distinguished in which the strength corresponds to the quantitative value presented in high-tech concrete and establishes the optimal parameters of their composition. The increased dispersion of cement S_{sp} = 450 to 550 m²/kg obtains concrete with a compressive strength of more than 75 MPa at 28 days. The increased consumption of PFM-1 ≥ 0.7% is needed to obtain cast and self-compacted concrete mixes with the required W/C values.

Based on the data obtained in Table 1.21 and the mathematical model (1.26), a nomogram was built (Figure 1.31).

Composite Ash-Containing Cements and Effective Concrete Based on Them 49

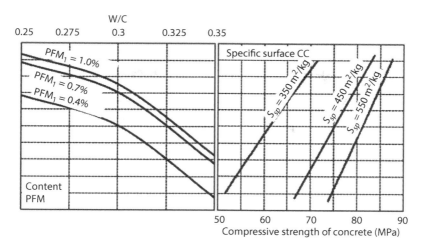

FIGURE 1.31 Nomogram for determining the strength of concrete based on ash CC with additives PFM at 28 days.

An important technological process that affects the structure and operational properties of concrete and reinforced concrete structures is heat treatment (HT). This technological stage in modern conditions is an effective method of accelerating hardening, entails a reduction in the production process and increase productivity. It largely determines the final physico-mechanical properties of concrete; the use of certain HT modes allows controlling the processes of structure formation. This is especially important for concrete, which has high demands on durability, for example, for road or hydrotechnical construction. In this case, "soft," low-temperature modes should be used, at the same time, enough stripping and transfer strengths can be achieved.

The aim of the study was to establish the optimal parameters of the HT regimes, as well as their influence on the kinetics of the set of strength and operational properties of concrete based on ash-containing CC. In order to achieve a required data with the minimum number of experiments, mathematical planning of the experiment was used [40] and the level of variations of the factors considered in this study are given in Tables 1.22 and 1.23.

After processing and statistical analysis of the experimental data, the concrete regression equations, of compressive strength were obtained:

After 4 hours HT:

$$f_C^{HT} = 58.3 - 0.46x_1 - 11.24x_2 + 1.36x_3 + 2.6x_4 + 0.8x_1x_2 \\ + 0.14x_1x_3 + 0.26x_2x_3 - 1.7x_1^2 + 4.0x_2^2 + 0.8x_3^2 + 1.85x_4^2; \quad (1.31)$$

TABLE 1.22
Experiment Planning Conditions

Factors		Variation Levels			
Natural	Coded	−1	0	+1	Interval
Content PFM$_1$ in CC (%)	x_1	0.4	0.7	1.0	0.3
W/C	x_2	0.25	0.3	0.35	0.05
Isothermal exposure time (τ_{ht}, h)	x_3	4	6	8	2
Steaming temperature (T_{is}, °C)	x_4	60	75	90	15

TABLE 1.23
Experimental Results of Concrete Strength at CC After HT

	Influence Factors				Compressive Strength (MPa)		
No	PFM$_1$ (%)	W/C	τ (hours)	T_{is} (°C)	4 Hours after HT	28 Days after HT	28 Days Normal Hardening
1	1.0	0.35	8	90	57.3	59.0	55.6
2	1.0	0.35	8	60	52.8	56.0	55.8
3	1.0	0.35	4	90	55.1	57.6	54.9
4	1.0	0.35	4	60	47.9	56.9	55.8
5	1.0	0.25	8	90	77.6	81.5	71.8
6	1.0	0.25	8	60	71.9	79.0	70.5
7	1.0	0.25	4	90	74.0	80.1	72.8
8	1.0	0.25	4	60	70.1	77.4	71.3
9	0.4	0.35	8	90	56.2	60.6	54.9
10	0.4	0.35	8	60	51.4	60.2	54.7
11	0.4	0.35	4	90	53.4	60.7	54.7
12	0.4	0.35	4	60	47.9	59.1	54.2
13	0.4	0.25	8	90	80.0	82.1	71.2
14	0.4	0.25	8	60	74.6	80.2	70.5
15	0.4	0.25	4	90	77.9	80.1	70.8
16	0.4	0.25	4	60	72.8	78.9	70.2
17	1.0	0.3	6	75	56.3	64.2	64.5
18	0.4	0.3	6	75	57.1	65.1	65
19	0.7	0.35	6	75	50.5	60.2	65.3
20	0.7	0.25	6	75	74.3	78.5	69.9
21	0.7	0.3	8	75	60.0	66.3	65.1
22	0.7	0.3	4	75	58.4	65.2	65.6
23	0.7	0.3	6	90	62.4	66.2	66.5
24	0.7	0.3	6	60	58.1	62.9	65.1

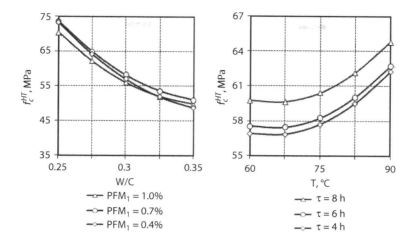

FIGURE 1.32 The influence of technological factors on the strength of concrete after steaming.

At the age of 28 days:

$$f_c^{HT,28} = 65.04 + 0.4x_1 - 10.5x_2 + 0.5x_3 + 0.97x_4 \\ - 0.49x_1x_2 + 0.24x_1x_4 - 0.3x_2x_3 + 0.16x_2x_4 \\ - 0.5x_1^2 + 4.2x_2^2 + 0.6x_3^2 - 0.6x_4^2. \quad (1.32)$$

The graphical dependences of the strength of steamed concrete based on ash CCs on technological factors 4 hours after HT are shown in Figure 132, and at the age of 28 days in Figure 1.33.

Their analysis allows placing the factors of influence on the strength of concrete after heat treatment in the following in row: $x_2 > x_4 > x_3 > x_1$. As expected, the W/C factor is the most influential; however, some interaction with the heat treatment parameters is observed—duration and maximum temperature. Among these two factors, temperature is more influential, an increase in the duration of isothermal exposure by 2 hours, allows to compensate for the decrease in the maximum temperature by 15°C (points 22 and 24), ceteris paribus.

In the case of an increase in isothermal exposure from 4 to 8 hours, is observed an increase in strength within 10% to 15%, as well as with an increase in maximum temperature by 25°C to 30°C. An increase in the amount of SP in the binder from 0.4% to 0.7% to 1.0% may be accompanied by a slight decrease in the strength of steamed concrete (Figure 1.32). However, this gives a mixture of cast consistency with a cone spread of up to 60 cm, which allows its use for concretes capable of self-compaction.

At the age of 28 days (Figure 1.33), the strength of concrete with different heat treatment modes is practically equalized. The strength of the samples of normal hardening is in all cases lower than that of steamed ones even under low-temperature

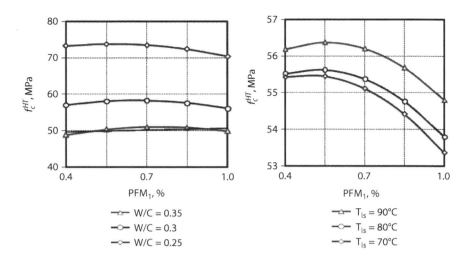

FIGURE 1.33 The influence of technological factors on the strength of steamed concrete at 28 days.

regimes of heat treatment. This lower strength can be explained by a more significant intensification of the hardening processes in the heat treatment of concrete based on ash-containing CC than concrete in ordinary Portland cement.

1.5.3 Deforming Properties

The features of the pore structure, the composition and structure of cement stone, and its volumetric concentration have a significant effect on the deforming properties of concrete.

The elastic modulus of concrete, E_c, can be expressed based on theoretical data obtained when considering concrete as a two-phase system with spherical particles uniformly distributed in a cement stone [56]:

$$E_p = \frac{V_{cs} + (2-V_{cs})E_{aggr}/E_{cs}}{2-V_{cs}+V_{cs}E_{aggr}/E_{cs}}, \qquad (1.33)$$

where V_{cs} is the volumetric concentration of cement stone, and E_{aggr} and E_{cs} are the elastic moduli of aggregate and cement stone.

The modulus of elasticity of a cement stone is decisively influenced by its porosity, V_p, and the elasticity of the so-called "gel-crystallite phase" (Eg), which follows from the well-known [61] expression:

$$E_{c.s} = (1-V_p)^3 E_g \qquad (1.34)$$

Available literature data on the influence of SAS are contradictory, although most researchers believe that the introduction of additives reduces the elastic modulus of concrete. The decrease in the elastic modulus with surfactant additives is explained by adsorption modification of the structure of cement stone and the appearance of a larger number of structural elements in hydrated shells of grains, which leads to an increase in the sliding surface of particles. When introducing plasticizing additives, along with modifying the structure, the influence of two possible opposite effects should be considered: a decrease in the porosity of cement stone and a decrease in the volume of cement stone in concrete.

V. Batrakov analyzed a number of experimental results on the effect of the additive SP-1 SP on the elastic modulus and came to the conclusion that for cast concrete the additives with a slight increase in strength (103.3%) leads to a slight decrease in the modulus elasticity (98.8%) [58]. When using SP to reduce W/C and increase strength, the modulus of elasticity of concrete can increase significantly. The elastic modulus of concrete is often associated with its compressive strength. The most common dependencies are:

$$E_\tau = \frac{E_m f_c^\tau}{S + f_c^\tau}, \quad (1.35)$$

where E_τ is the modulus of elasticity of concrete under load at the age τ, f_c^τ is compressive concrete strength at the same age, and E_m and S are empirical coefficients ($E_m = 2000$ and $S = 23$).

An analysis of the previous formula shows that the use of CC can lead mainly to an increase in the elastic modulus depending on the nature of the change in structural parameters and the strength of concrete.

Determination of the modulus of elasticity and prismatic strength of high-tech concrete was carried out at 28 days on prisms measuring 70–70–280 mm. Deformations were measured using strain gauges with an accuracy of 0.7×10^{-5}. The prism load was created in spring units in steps of 0.05 R_{pr} aging of 5 minutes to the level of 0.3 R_{pr} (R_{pr} is the prismatic strength). The value of the elastic modulus was calculated by the formula:

$$E_c = \sigma / \varepsilon_{pr} \quad (1.36)$$

where σ is the applied stress and ε_{pr} is the longitudinal relative deformation.

As follows from the data Table 1.24, the ratio of the prismatic and cubic strength of concrete based on ash-containing CC is in the range of 0.75 to 0.77, which corresponds to the well-known correlation dependence.

The experimental values of the elastic modulus of the studied concrete are slightly higher than calculated by the formula (1.35); it does not require specification of the coefficients in this dependence. As for ordinary concrete, the important influence on the nature of the relationship between the modulus of elasticity and the strength of concrete that has been shown by many studies [46], have the elastic properties of the aggregate, W/C, and the content of the cement paste in the concrete mixture.

TABLE 1.24
Deformative Properties of Concrete Based on CC

No.	The Composition of the Concrete Mixture			Strength, 28 Days (MPa)		Elastic Modulus, E_c MPa × 10^{-4}		Deformation Shrinkage, ε_{shr} mm/m × 10^{-5} in, days	
	Content of CC (kg/m³)	S_{sp} (m²/kg)	W/C	Cubic	Prismatic	Estimated	Experimental	28	90
1	500	450	0.25	70.1	53.5	5.50	5.55	1.7	2.1
2	500	550	0.25	74.9	58	5.78	5.83	2.1	2.4
3	500	450	0.35	54.7	41.5	4.55	4.61	2.4	2.6
4	500	550	0.35	59.8	45.7	4.88	4.94	2.6	2.8
5	300	450	0.35	41.4	31.1	3.64	3.68	2.1	2.4
6	300	550	0.35	45.4	34.5	3.92	3.95	2.3	2.5

The shrinkage deformations of the studied concrete (Table 1.24) were determined using dial gauges with a division value of 1 μm installed on two opposite faces of prisms based on 200 mm. The experiments were carried out at a temperature of 18°C ± 2°C and relative humidity 75% ± 5%.

In addition to water content, shrinkage also is affected, although to a lesser extent, by the value of W/C, specific surface area of cement, and its content in concrete mix. The shrinkage of concrete based on CC is 1.2 to 2 times less than the shrinkage of concrete based on marketable cements. The obtained results (Table 1.24) show that when using CC with a specific surface area of 450 to 550 m²/kg, almost complete compensation of the shrinkage deformation of concrete is possible. The effect of reducing shrinkage deformations can be explained by the high water-retention ability of CC, transition of a significant amount of water to an adsorption-bound state, and the prevention of rapid surface drying of concrete.

1.5.4 CRACK RESISTANCE

The crack resistance of concrete depends on a whole range of factors: thermal expansion, creep, changes in the elastic-plastic characteristics of concrete with a change in temperature, etc. Therefore, the selection of the criterion and assessment of crack resistance of concrete is a rather difficult task. A decrease in the risk of cracking of concrete with addition of active mineral fillers, specifically fly ash, into its composition or composition of a binder is associated with a decrease in heat release [4]. This criterion is relevant for massive concrete. To assess the crack resistance of concrete in structures is most expedient to use another criterion—the ratio of splitting tensile strength to compressive strength ($f_{c.sp}/f_c$) [59]. An increase in this ratio is considered more favorable from the point of view of crack resistance as was shown in several studies [15,43].

Based on the data obtained (Table 1.21), a mathematical model and graphical dependencies characterizing crack resistance were obtained:

$$\frac{f_{c,sp}}{f_c} = 0.06 + 0.0017x_1 + 0.0027x_3 - 0.0003x_1 x_3 \\ -0.0001x_1^2 + 0.0014x_2^2 + 0.0009x_3^2. \quad (1.37)$$

An analysis of the mathematical model (1.37) and graphical dependences allows considering that for ash-containing concrete with additives PFM the ratio $f_{c.sp}/f_c$ is higher in comparison with concretes on non-admixtures cements. (Figure 1.34).

The decrease in the ratio $f_{c.sp}/f_c$, in time reflects the features of the processes of structure formation of cement stone, the composition of concrete, and the conditions of its hardening [65,66]. The situation known for ordinary cement concrete can be attributed fully to high-tech concrete based on ash-containing CC. For example, the variation of the content of the ash and PFM in the CC stabilizes $f_{c.sp}/f_c$ in the cube in a later period of 30–60 days. An interesting fact is that with an increase in the age of such concrete, the number of factors and their interactions that affect the ratio $f_{c.sp}/f_c$ decreases.

1.5.5 Water Impermeability

The water impermeability of concrete to a large extent depends on the features of its porous structure. If the total porosity has the main effect on the strength of concrete, then water impermeability is a function of open porosity. Several researchers [53,59] indicated that the main routes of water penetration into concrete are pores of sedimentary origin, the formation of which is most characteristic of concrete with increased water content. Sedimentation depends mainly on the viscosity of the cement paste and the main sedimentation characteristic of a concrete mix is its water separation.

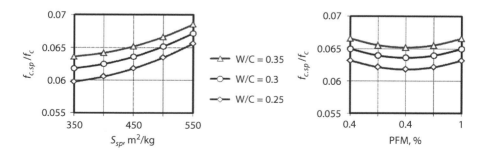

FIGURE 1.34 The influence of technological factors on the value of the coefficient of crack resistance of concrete at 28 days based on ash-bearing CC.

TABLE 1.25
Properties That Determine Durability High-Tech Concrete Based on CC

	The Composition of the Concrete Mixture			Compressive Strength, 28 Days		Water Impermeability,	Frost Resistance,
No.	Content of CC (kg/m³)	S_{sp} (m²/kg)	W/C	(MPa)	f_{pr}/f_c^*	W (MPa)	F, cycles
1	500	450	0.25	70.1	0.76	1.3	430
2	500	550	0.25	74.9	0.77	1.4	470
3	500	450	0.35	54.7	0.76	1.1	360
4	500	550	0.35	59.8	0.76	1.2	390
5	300	450	0.35	41.4	0.75	0.8	340
6	300	550	0.35	45.4	0.76	1	380

* f_{pr} is the prismatic strength.

As noted earlier, fly ash has a significant effect on the viscosity of plasticized cement paste and the water separation of the concrete mix. Due to fly ash, which is included in the CC and the increased dispersion of the binder in concrete, the number of micropores is reduced and open pores are clogged. The main factor determining the water permeability of high-tech concrete, as for traditional concrete, is the water-cement ratio. A slight increase in the water impermeability of concrete based on CC, as shown by the results of experimental studies (Table 1.25) with W/C from 0.25 to 0.4, can be explained by a decrease in the total porosity and a redistribution of the ratio between different types of pores.

With the increasing age of concrete, water impermeability increases as a result of irreversible changes in the structure of the pore space and an increase in the volume of the solid phase in the cement stone.

With a sufficient hardening moisture regime, the increase in concrete water impermeability in the later stages is much larger than the relative increase in compressive strength.

To determine the effect of the degree of dispersion of CC on the permeability, the nature of the porosity of concrete samples based on ash-containing CC of various dispersion was determined. It was confirmed that the introduction of CC with a dispersion of 350 m²/kg slightly reduces the number of open capillary pores in concrete. With an increase in the dispersion of the binder to S_{sp} = 450 to 550 m²/kg, the total number of capillary pores significantly decreases. However, high dispersion leads to a significant decrease in the equivalent pore radius and an increase in the specific surface area of the pores, which leads to a decrease in the permeability of concrete.

1.5.6 FROST RESISTANCE

The frost resistance of concrete is largely determined by the nature of its pore structure. Capillary pores are the main defect in the structure of densely laid concrete, which reduces its frost resistance [60,61]. According to G. Gorchakov, concrete with a capillary pore content of not more than 5% to 7% is frost-resistant. In addition, pore sizes have a significant effect on frost resistance; pores with a size of more than 5 to 10 cm negatively affect it.

Our results indicate a low open porosity of less than 3% to 5% of concretes based on finely dispersed CC and explain the possible increase in their frost resistance in comparison with traditional concretes. The critical numbers of freeze and thaw cycles are given in Table 1.25. Fly ash, which often leads to a decrease in frost resistance when introduced into the composition of CC with PFM additives, contributes to some increase in the frost resistance of high-tech concrete that is result of a decrease in pore size. An increase in the dispersion of the binder to $S_{sp} = 450$ to 550 m²/kg to a greater extent contributes to a decrease in the size of open capillary pores and, as a result, to an increase in the frost resistance of concrete based on ash containing CC.

1.5.7 Designing Compositions of High-Tech Concrete

The design and optimization of high-tech concrete compositions is to determining the ratio of components that provides a set of design indicators, specifically strength at different ages, with necessary workability of the concrete mix [62]. Since all strength characteristics can be unambiguously associated with the W/C, the problem can be solved by the following algorithm.

1. From the models of strength under normal hardening conditions (1.27) and after heat treatment (1.29 and 1.30), the W/C is determined, which provides each of the specified strength properties.

    To solve these models with respect to W/C, it is necessary to set the values of other factors, specifically the content of the PFM and the parameters of the heat treatment. In a first approximation, from the condition of minimum cost for concretes with moderately moving mixtures, take the minimum PFM content (0.4%). For cast mixtures, the PFM content should be increased to 0.7%, for concretes capable of self-compaction increase to 1.0%. Heat treatment parameters are accepted at a minimum level.
2. Certain values of W/C take the minimum, which provides all the specified properties. With significant (more than 25%) differences between the *W/C* values obtained under different conditions, the content of PFM and ash in the CC, the parameters of the heat treatment are changed to reduce the "scissors" of the W/C. For preliminary selection of composition parameters, nomograms can be used (Figures 1.28 through 1.32).
3. Determine the water consumption for a given workability from the model (1.25), or using the nomogram (Figure 1.27), determine the water demand of the concrete mix, depending on the initial parameters (specific surface area and cement consumption, and PFM content).
4. Determine the consumption of cement in kilograms per cubic meter:

$$C = W / (W/C) \qquad (1.38)$$

TABLE 1.26
Correction for Water Consumption of Concrete Mixture at a Flow Rate Cement (Consumption Greater than 400 kg/m³)

Specific Surface Area CC (m²/kg)	Composite Cement Consumption (kg/m³)								
	420	440	460	480	500	520	540	560	580	600
	Increased Water Demand of the Concrete Mix to (versus initial water) Calculated (kg/m³)								
550	4	8	12	15	20	25	30	35	40	45
450	3	6	9	12	15	19	23	27	31	35
350	2	4	6	8	10	12	15	18	21	25

If the resulting cement consumption exceeds the maximum possible value, the PFM consumption should be increased by listing it according to equation (1.25) or the nomogram by Figure 1.27. The correction for the increase in water consumption is taken according to Table 1.26. If a decision is made to increase water demand, then the cement consumption should be listed considering the adjusted water demand.

5. Determine the consumption of aggregates by the method of absolute volumes.
6. Take the specified process parameters (cement dispersion, PFM content, and HT parameters) at acceptable levels and perform technical and economic optimization of concrete composition at cost.

1.5.8 Example of Calculating the Composition of Concrete Type HPC

It is necessary to design a concrete composition of type HPC class C50/60, with slump cone, $S = 22$ cm, using CC (fly ash 40%) with specific surface area $S_{sp} = 450$ m²/kg. Granite crushed stone 5–20 mm with $\rho_{cr.st} = 2700$ kg/m³ and $\rho^b_{cr.st} = 1450$ kg/m³, sand of medium size with $\rho_s = 2650$ kg/m³ and $\rho^b_s = 1470$ kg/m³.

The average strength of concrete of class C50/60, which is determined on cubic samples with a coefficient of variation of 13.5%, is 77 MPa. Only compressive strength is normalized; therefore, the nomogram (Figure 1.31) is used to determine the necessary W/C = 0.315.

1. With the initial parameters of mobility $S = 22$ cm and dispersion of cement $S_{sp} = 450$ m²/kg, as a first approximation, the approximate cement consumption was set as $C = 400$ kg/m³ and additive PFM = 0.7%. Using the nomogram (Figure 1.27), the consumption of water, $W = 145$ L/m³, was determined.

   Determine the consumption of cement in kilograms per cubic meter:

   $$C = 145 / (0.315) = 460 \text{ kg.}$$

2. According to Table 1.26, the water consumption was adjusted at $C = 460$ increasing the amount of water by 9 L. Thus, $W = 154$ L.
3. The final calculated cement consumption in kilograms per cubic meter:

$$C = 154/0.315 = 489 \text{ kg.}$$

4. Determine the consumption of aggregates by the method of absolute volumes:
   Crushed stone (kg/m³):

$$Cr.st = \frac{1000}{\alpha \dfrac{V_{cr.st}^p}{\rho_{cr.st.}} + 1 / \rho_{cr.st}} = \frac{1000}{1.37 \dfrac{0.46}{1.45} + \dfrac{1}{2.7}} = 1238,$$

where α is the grain partition coefficient and $V_{cr.st}^p$ is the voidness of coarse aggregate

$$V_{cr.st}^p = 1 - \frac{\rho_{cr.st}}{\rho_{cr.st}} = 1 - \frac{1.45}{2.7} = 0.46$$

Sand (kg/m³):

$$S = \left(1000 - \left(\frac{C}{\rho_c} + W + \frac{Cr.st}{\rho_{cr.st}}\right)\right) \cdot \rho_s$$

$$= \left(1000 - \left(\frac{489}{3.1} + 145 + \frac{1238}{2.7}\right)\right) \cdot 2.65 = 609.$$

Estimated nominal concrete composition (kg/m³): $C = 489$ kg/m³; at W/C = 0.315; $W = 154$ kg/m³; PFM $= 0.007 \times 489 = 3.42$ kg/m³; $S = 609$ kg/m³; $Cr.st = 1238$ kg/m³.

2 Activated Low Clinker Slag Portland Cement and Concrete on Its Basis

2.1 ACTIVATION OF SLAG BINDERS

2.1.1 METHODS FOR ACTIVATING SLAG BINDERS

The development of modern construction technologies in all technically advanced countries is aimed at the development of efficient materials, the use of which is economically feasible, reduces energy costs, and reduces consumption of raw materials [63]. In concrete, as a rule, cement accounts for 50% to 70% of the total cost of energy resources. Therefore, the problem of reducing the energy intensity of cement production is one of the main tasks of the cement industry [64].

The production of 1 ton of Portland cement requires 1.3 to 1.6 tons of natural raw materials, and the cost of fuel and electricity in the clinker cost is 60% to 70%. This problem is of relevance for the wet cement production when the average fuel consumption reaches an average of 220 to 230 kg/t clinker.

One of the real ways to reduce energy consumption is to increase the volume of production of composite, including low clinker cements.

Each ton of clinker saved in cement will save 3,500 to 7,500 MJ of energy. At the same time, 0.8 to 1.2 tons of CO_2 are emitted from one ton of cement, so reducing clinker production leads to an equivalent reduction in CO_2 emissions [65,66].

Low-clinker cements filled with industrial wastes (slag, fly ash, etc.) have several positive features: low cost, less energy intensive production process, recycling of accumulated waste, and reduction of harmful emissions into the atmosphere. However, such cements are not yet very popular for manufacturers mainly because of their relatively low strength. Such cements are characterized by slow hardening speed and increased water demand [67].

In accordance with European standard for cement, EN 197-1 allowed production of slag Portland cement with a clinker content of 5% to 19%. In this regard, there is a need to find effective technical solutions to improve the effectiveness of low clinker cements. Analysis of the results of research and practical experience in the development of cement with a high content of mineral additives, as well as known patterns of structure formation of multicomponent binder systems, shows the manufacture of a new generation of low clinker cements with improved construction and technical properties is possible by finer grinding and mixing with complex chemical additives of polyfunctional action.

Even with the most favorable chemical composition, blast furnace slags at normal temperatures (up to 15°C to 25°C) have almost no activity when interacting with water. However, as shown in numerous studies, blast furnace slags have crystalline and vitreous phases that are capable under the influence of mechanical, chemical, and thermal factors to interaction with water and hydraulic hardening, which is due to the formation of new insoluble water substances. Various authors demonstrate this capability clearly for the many varieties of blast furnace slag. Studies have established the dependences of their activity on the content of different minerals, microfillers, and additives of activators [19].

The interaction of blast furnace slag with water is a complex of processes including adsorption, ion exchange, hydration, hydrolysis, leaching, etc., which results in the destruction of the starting material and the emergence of new phases. The breaking of the bonds (first, high-ionic Me-O, then more covalent Si-O-Si and Al-O-Si bonds) leads to depolymerization of the silicon tetrahedra of the original structure. The intensity of the curing processes, the structure, and composition of the neoplasms depend on the curing conditions and the type of slag activator used.

Increasing the temperature and pressure of the aqueous medium greatly changes its properties, which determine the course of hydrolytic degradation of slag crystalline and vitreous phases, specifically the degree of dissociation of water, the mobility of H^+ and OH^- ions, pH, and viscosity. The degree of dissociation of water increases with increasing temperature and pressure.

The higher mobility of protons determines their ability to accelerate reactions in silicate systems and act as catalysts. Increasing the mobility of H^+ and OH^- ions with increasing temperature and pressure of the aqueous medium, along with increasing their number, explains the increase in electrical conductivity.

The pH of pure water at equilibrium vapor pressure decreases with increasing temperature (6.99, 6.14, 5.60, respectively, at 25°C, 100°C, and 250°C) [68]. The breakage of hydrogen bonds that occurs in water as the temperature and pressure increase will reduce its viscosity and increase the penetration of the porous body. The solubility of water increases with increasing temperature due to the destruction of its frame.

As is known [68], the water solubility of amorphous, crystalline, and glassy silica increases with temperature. Thus, as a result of the destructive action of the hydrothermal medium on the grouping of silicon and alumina tetrahedra, the processes of hydrolytic destruction of crystalline and vitreous phases of the slag are accelerated at elevated temperatures and pressure, thus their solidification occurs.

According to known data including data herein, the steamed beams of the composition of the binder:sand = 1:3, made by mixing of sand and ground blast furnace slag with pure water, are characterized by a compressive strength of 5 to 10 MPa. The products of hardening of slag in water are low-basic hydrosilicates, hydroaluminates, and calcium hydroferites. However, the level of strength of the slag binder thus obtained is very low and of no interest to the builders and the construction industry.

In the presence of Ca^{2+} and SO_4^{2-} ions, the hydration of both the crystalline and glassy phases of blast furnace slag is accelerated. Moreover, when hardening occurs under normal conditions, the presence of both ions is necessary, and in the conditions of heat treatment, the presence of excess Ca^{2+} is enough [39]. At room temperature, in the presence of Ca^{2+} and SO_4^{2-} ions, the minerals gelenite and okermanite are actively hydrated, but other slag minerals require hydrothermal treatment.

The glassy phase of the slag in the presence of lime reacts with water more actively than the crystalline phase. This result is because at ordinary temperatures the glass is an unstable phase; however, under the influence of external factors (in this case, water and lime) the glass tries to go into a stable crystalline phase. High internal chemical energy provides increased solubility of glass, the result of which is the formation of metastable supersaturated solutions and their crystallization.

The acceleration of the hydration process of vitreous slag in the presence of Ca^{2+} ions in water is caused by the destruction of the shell of $Al(OH)_3$ and $Si(OH)_4$ on hydrated glass grains as a result of the interaction of these hydroxides with Ca^{2+} to form low-basic hydrosilicates and hydroaluminates. In this case, the crystallization of the shell is accompanied by the formation of a network of capillaries through which water molecules reach the non-hydrated deep region of the particles of glass. If free Ca^{2+} ions are present in the solution, the formation of new continuous shells on the slag glass particles does not occur, which ensures a slow but continuous flow of hydration processes [39].

CaO, $Ca(OH)_2$, $Mg(OH)_2$, and cement clinker are used as alkaline-earth activators of slag hardening.

The products of the hydration of the glassy and crystalline phases of slag in the presence of activators include CSH, C_2SH, C_3S_2H, hydrogelenite, and hydrogranates.

V. D. Glukhovsky developed the physico-chemical, basis to produce slag alkali binders (SAB). The composition of the products of neoplasms of SAB is determined by the composition of slag, the type and concentration of the solution of the alkaline component, and the curing conditions. The hardening of SAB takes place in three stages [7]:

Stage 1. Adsorption-chemical dispersion of slag.
Stage 2. Formation of a dispersion-coagulation structure of the binder system due to the increase in the number and area of contact between the particles in the process of adsorption-chemical dispersion of the slag.
Stage 3. Crystallization of neoplasms and formation of crystalline structure of binder stone.

All types of blast furnace slag—acidic, neutral, basic, as well as steel, electro-thermo-phosphoric, non-ferrous metallurgy slags—can be used as for the manufacture of SAB. As alkaline components, it is most economically feasible to use alkaline wastes of various types of industry: soda-sulfate mixture, alkali melting, sodium aluminates, etc.

The slag-alkali binders harden under normal conditions and during heat treatment; their strength depends on the type and concentration of the solution of the alkaline component. The binder strength reaches 100 to 50 MPa when using solutions of meta and disilicate sodium and 40 to 60 MPa when using solutions of alkaline carbonates and caustic alkalis.

The production and use of SAB have several problems:

1. Sodium meta- and disilicate, which are required to produce high strength binders, are relatively expensive and not available to many manufacturers.
2. Some types of SAB are characterized by very short setting time (especially when using meta and sodium disilicate).
3. In concrete and reinforced concrete structures based on SAB, under certain conditions, efflorescences from chemically unbound alkalis may appear.

Hydration of slag is possible under the influence of not only alkaline but also sulfate activator. In the presence of SO_4^{2-} (and also Ca^{2+}) ions in the curing system of slag-water, crystals of calcium hydrosulfoaluminates are formed, which prevent the formation of watertight shells of aluminum hydroxides and silicon on glass particles, and in the case of early formation of such shells, promote their destruction. This process also intensifies the ion exchange of $Me^+\leftrightarrow H^+$ on the surface layer of particles, which causes deformation of the structure of the glass. In solution in the presence of SO_4^{2-}, needle crystals of hydrosulfoaluminate calcium [39] are formed, which contribute to the hardening. At the same time, the formation of low-basic calcium hydrosilicates occurs; according to [7,19,39], sulfate activation is most effective for alumina blast furnace slag and alkaline is suitable for slag of all kinds.

When studying the effect of different activators on the properties of blast furnace slag, it is established that as activators of hydration processes, various modifications of calcium sulfate can be used ($CaSO_4 \cdot 2H_2O$, $CaSO_4 \cdot 0.5H_2O$, insoluble or soluble anhydrite, phosphogypsum (FG), etc.) in the presence of a small amount of alkali or of the substances forming them (Portland cement clinker, burnt dolomite, lime).

With double activation of slag with calcium hydroxide and sulfate component, along with the positive sides, destructive processes can occur under certain conditions—slowing of hardening and even falling of strength. This result is due to the fact that under certain conditions in the CaO-Al_2O_3-$CaSO_4$-H_2O system recrystallization processes are possible due to the formation of compounds that cause an increase in the volume of the solid phase in the hardened artificial stone that leads to the concentration of local stress, decrease of strength, and even destruction. Therefore, it is necessary to consider the optimal quantitative ratio of the activators used.

2.1.2 Activation of the Low Clinker Slag Portland Cement (LSC)

One of the main ways of activating an LSC is to increase its specific surface area. Several experiments have compared the effectiveness of single-stage and two-stage grinding of LSC, as well as the effect of the grinding intensifier propylene glycol and superplasticizer additive Sika VC 225 on the kinetics of grinding. The results of the studies are given in Table 2.1 and Figure 2.1.

As a binder in studies, an LSC with a clinker content of 12% and FG 7.5% (4.5% in terms of SO_3) was used. The binder grinding was carried out in a laboratory ball mill. In one-stage grinding, there was a compatible grinding of all binder components, and in a two-stage grinding, first the cement clinker was ground separately to a specific surface of 250 m^2/kg followed by a compatible grinding of all binder components.

TABLE 2.1
Influence of Grinding Duration and Method of Introduction of Additives on Strength of LSC

			One-Stage Grinding				Two-Stage Grinding			
	Additive	Duration			Compressive Strength (MPa)				Compressive Strength (MPa)	
No.	Type	Grinding (hours)	Specific Surface (m²/kg)	Normal Consistency (%)	7 Days	28 Days	Specific Surface (m²/kg)	Normal Consistency (%)	7 Days	28 Days
1	Without additives (control)	1.5	322	26.5	7.3	21.1	301	24	10.2	29.4
		2	366	25	9.2	26.0	355	22	11.8	31.0
		3	454	25	12.2	31.3	453	23	15.6	41.9
2	Propylene glycol (0.05%)	1.5	356	25.5	11.0	23.9	358	24.5	19.2	30.1
		2	451	25	13.6	34.4	457	24.5	21.2	42.5
		3	559	31	14.8	32.0	554	29	22.2	36.0
3	Sika VC 225 (0.3%) when grinding	1.5	311	20	6.5	13.8	300	19	19.6	34.6
		2	356	18.5	10.2	19.4	356	17	25.2	38.0
		3	443	17	15.6	30.1	450	17	28.6	41.4
4	Sika VC 225 (0.3%) when mixing	1.5	322	26.5	7.6	22.4	301	24	10.8	39.6
		2	366	25	9.0	27.0	355	22	14.8	48.1
		3	454	25	12.2	36.4	453	23	19.5	55.2
5	SP-1 (0.5%) when grinding	1.5	318	25	8.5	20.9	311	23	9.3	29.5
		2	369	23	11.6	26.5	360	21	13.8	32.2
		3	460	23.5	13.0	33.3	457	22	22.5	43.4
6	SP-1 (0.5%) when mixing	1.5	322	26.5	9.8	23.3	301	24	10.5	30.8
		2	366	25	11.6	27.2	355	22	16.3	39.2
		3	454	25	14.8	33.5	453	23	22.8	45.0

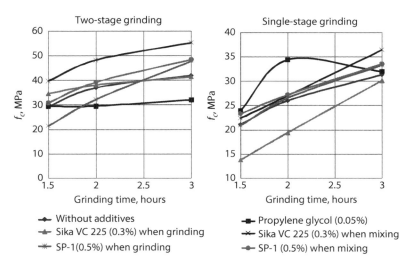

FIGURE 2.1 The influence of the duration of grinding a binder on its activity.

The results obtained indicate that the specific surface area is a significant factor affecting the activity of the LSC. With an increase from 300 to 320 m²/kg to 450 m²/kg, activity increases on average of 25% to 30%. A more efficient two-stage grinding method was found in which the first grinding was of cement clinker followed by a joint grinding of clinker and slag. In this case, the activity of the binder is increased by 30% compared to one-stage grinding. These results coincide with the data [22,33] that with two-stage grinding, it is possible to significantly reduce the size of clinker grains and significantly increase the activity of the blast-furnace cement.

Studies used an LSC with a clinker content of 12% and FG 7.5% (4.5% in terms of SO_3) as a binder. The binder grinding was carried out in a laboratory ball mill. In one-stage grinding, there was a compatible grinding of all binder components, and in a two-stage grinding, first the cement clinker was ground separately to a specific surface of 250 m²/kg followed by a compatible grinding of all binder components.

To reduce energy consumption in our studies, protylele glycol (PG) was used as an additive grinding intensifier. An increase PG in the content of the binder mass to a value of 0.05% allows a reduction in the grinding time (Figure 2.1) and, accordingly, a reduction in energy consumption by one-third, and this also increases in the early strength of the samples, which is associated with an increase in the amount of small particles in the binder. The introduction of additives—superplasticizers, Sika VC 225, and SP-1—does not affect significantly the kinetics of binder grinding. A positive effect is observed with the introduction of these additives during the mixing of a cement-sand mortar with a composition of 1:3. At the same time, the strength of the samples increases on average by 25% when using Sika VC 225 superplasticizer and by 15% when using SP-1.

An increase in the fineness of grinding LSC can be considered a method of its mechanical activation. Along with it, the chemical activation of LSC is of practical interest. This work studies the effectiveness of sulfate and complex *sulfate-fluoride-alkaline activation* of LSC.

For studies of sulfate activation of LSC, FG was used and, for comparison, gypsum stone.

FG is a waste product of processing apatite concentrate with sulfuric acid to produce phosphoric acid. Under industrial conditions, the process of decomposition of calcium fluoroapatite is carried out at 65°C to 80°C. The resulting product contains phosphate anhydrite and sludge ($CaSO_4 \cdot 2H_2O$). The yield of FG waste (on dry matter) is 4 to 5 tons per 1 ton of P_2O_5. According to electron microscopic studies [5,19], the bulk of the particles of FG is represented by well-formed crystals of gypsum up to 1 to 1.25 mm long (mainly 0.05 to 0.2 mm).

A binder with a clinker content of 12% was used for these studies. The content of the sulfate component varied from 3.5% to 5.5% in SO_3 with an interval of 1%. The binder was milled according to a two-stage grinding until reaching a specific surface of 450 m²/kg. To reduce water demand, an additive of superplasticizer SP-1 was used in an amount of 1% by weight of the binder. The study results of sulfate activation of LSC are presented in Table 2.2.

According to the data obtained, FG is the most effective sulfate component among the two used. A probable reason for this result is the structural features of FG, which is characterized by the porosity particles continuing even after grinding [66], the high dispersion, as well as the presence of fluorite impurities, mainly CaF_2, in it. The latter is also an activator of slag hardening in binders [20]. In addition, FG is also the cheapest sulfate component; its use contributes to solving the environmental problems of storage and disposal of potentially hazardous industrial waste. However, an increase in the content of the sulfate component of more than 7.5%, when using FG and gypsum stone, leads to a decrease in the strength of the LSC. This result is due to the fact that during the double activation of slags by calcium hydroxide—formed as a result of clinker hydration—and the sulfate component, destructive processes as a result of recrystallization of hydrosulfoaluminates are also possible, thus leading to a slowdown in hardening and a certain decrease in strength.

To non-calcined sulfate, activators can be attributed to common and relatively inexpensive water-soluble sulfates of various metals—Na_2SO_4, $MgSO_4$, and $Fe_2(SO_4)_3$. The results of studies of LSC with additives of water-soluble sulfates are presented in Table 2.3.

According to the data obtained, metal sulfates do not significantly affect the strength of LSC with a specific surface of 450 m²/kg, and with an increase in their content over 1%; a negative effect is even observed. This effect can be explained by the fact that an increase in the content of the sulfate component leads to a deficiency of Ca^{2+} ions in the binder system necessary for the formation of hydrosulfoaluminates and results in a decrease in binder strength. This conclusion also is confirmed by the appearance of many efflorescences during prolonged storage on the surface of the samples with an increase in the content of metal sulfates [7]. However, some positive effect of sulfates on the strength of LSC is observed when the specific surface area of the binder is reduced to 350 m²/kg. Moreover, the introduction of the additives Na_2SO_4 and $Fe_2(SO_4)_3$ in an amount up to 1% leads to an increase in strength by about 10% to 15%.

The main factor affecting the strength of LSC, as in previous studies, remains the specific surface area with an increase in which from 320 to 350 m²/kg and 400 to 450 m²/kg, almost doubles the strength.

TABLE 2.2
The Effect of the Type of Sulfate Component on LSC Properties

| No. | The Content of Components (%) | | | | Normal Consistency (%) | Setting Time (hours) | | Bending Strength (MPa) | | Compressive Strength (MPa) | |
	Slag	Clinker	Phosphogypsum According to SO$_3$	Gypsum Stone (by SO$_3$)		Initial	Final	7 Days	28 Days	7 Days	28 Days
1	88	12	3.5	—	22.7	3.5	6.5	6.8	10.8	13.4	34.1
2	88	12	4.5	—	23.3	3.2	7.5	7.4	10.8	15.6	41.9
3	88	12	5.5	—	24.5	4	8	5.3	9.8	14.5	36.2
4	88	12	—	3.5	23.5	3.9	7.4	5.5	8.2	12.3	30.8
5	88	12	—	5.5	25.5	3.5	8.6	5.9	10.5	25.8	33.5
6	88	12	—	5.5	26.2	4.5	9.0	4.5	7.6	13.2	31.5

TABLE 2.3
Study of the Effectiveness of Sulfates of Various Metals as Sulfate Activators in LSC

Additive			Specific Surface Area 350 m²/kg					Specific Surface Area 450 m²/kg			
			Bending Strength (MPa)		Compressive Strength (MPa)			Bending Strength (MPa)		Compressive Strength (MPa)	
No.	Type	Content (%)	W/C	7 Days	28 Days	7 Days	28 Days	W/C	7 Days	28 Days	7 Days	28 Days
1	No additives	—	0.33	5.1	9.1	13.6	34.4	0.32	7.4	10.8	21.2	42.4
2	Na$_2$SO$_4$	1.0	0.36	5.6	9.6	20.2	40.8	0.34	8.6	9.8	26.6	42.5
3	Na$_2$SO$_4$	2.0	0.36	5.0	8.4	16.2	36.0	0.36	7.4	11.1	26.2	40.5
4	Fe$_2$(SO$_4$)$_3$	1.0	0.34	5.4	8.9	17.6	40.5	0.34	7.7	10.1	27.6	39.7
5	Fe$_2$(SO$_4$)$_3$	2.0	0.36	5.2	8.8	13.4	21.8	0.36	6.5	11.2	24.2	38.1
6	MgSO$_4$	1.0	0.34	4.5	7.3	12.4	23.4	0.34	7.1	10.7	24.4	38.5
7	MgSO$_4$	2.0	0.34	4.7	7.9	13	23.0	0.36	5.1	9.4	15.4	37.0

Note: The content of FG in LSC is 7.5% by weight (in terms of SO$_3$, it is 4.5%)

Along with sulfate activators, this work studies the possibility of obtaining an additional activation effect when using other salt additives. For this purpose, additives of sodium silicon fluoride, calcium chloride and iron, as well as sodium carbonate were used. As a control, binder used LSC with a specific surface of 350 and 450 m²/kg (composition: Slag, 88%; clinker, 12%; and FG, 7.5% (above 100%). To reduce the water consumption necessary for the mortar to reach the standard cone spreading value of 106 to 115 mm, at all points during mixing the superplasticizer SP-1 was added in an amount of 1% by weight of the binder.

Studies have shown that the most effective hardening accelerator for LSC with a specific surface area of $S_{sp} = 350$ m²/kg, is sodium silicofluoride (Na_2SiF_6), which belongs to the group of fluorite salts (Table 2.4).

A characteristic feature of these salts is their low solubility in water. According to [7], the solubility of Na_2SiF_6 is 0.6 g/100 g of water at 18°C. When fluorite salts are used, protonation of water molecules adsorbed on the surface of the slag particles occurs by replacing the OH group with an F⁻ ion. At the same time, the activity of the binder increases, and hydration reactions are activated [20].

Ion F⁻ performs activating influences in three directions:

1. Activates the breaking of Si-O bonds and the transfer of silicon into solution.
2. Replaces OH-groups on the surface of particles with F.
3. Changes the effective charge on the atom of the active center (Ca or Si), strengthens the bonds during the formation of $Ca(OH)_2$ and $Si(OH)_2$ complexes, strongly deforms the water molecule, promotes its dissociation and the formation of active H⁺ and OH ions.

According to [7], the ionic salts CaF_2, MgF_2, and Na_2SiF_6 have an activating effect on the hardening of slags and slag Portland cement. The influence of fluorites on the activity of surface centers is carried out by ion absorption, while the OH⁻ group is substituted by the F-ion, which is more electronegative.

The strength of LSC with the addition of Na_2SiF_6 increased by more than 1.5 times compared to the strength of the control binder. The stronger effect of Na_2SiF_6 on the strength of the LSC, compared with other additives, can be explained by the appearance of significant amounts of $Si(OH)_4$ gel during the hydrolysis of Na_2SiF_6 in water by reaction:

$$Na_2SiF_6 + 4H_2O = Si(OH)_4 + 2NaF + 4HF.$$

The resulting Na_2SiF_6 gel actively interacts with LSC with the formation of low basic hydrosilicates. Hydrofluoric acid is highly soluble in water (the concentration of aqueous solutions reaches more than 50% [7]) and actively interacts with slag particles.

Thus, the use of a relatively inexpensive and common salt of Na_2SiF_6 significantly (by 35% to 40%) increases the strength of LSC with a moderate consumption of additives and the specific surface area of the binder.

TABLE 2.4
The Effect of Additives of Hardening Accelerators on the Strength of LSC

No.	Additive Type	Content (%)	Specific Surface Area 350 m²/kg						Specific Surface, 450 m²/kg					
			W/C	Bending Strength (MPa)		Compressive Strength (MPa)		W/C	Bending Strength (MPa)		Compressive Strength (MPa)	
				7 Days	28 Days	7 Days	28 Days		7 Days	28 Days	7 Days	28 Days
1	No additives (control)	—	0.33	5.1	9.1	16.3	33.1	0.32	7.4	10.8	23.4	45.3
2	Na_2SiF_6	1.0	0.35	7.1	9.5	28.4	33.2	0.34	7.5	10.0	28.7	46.1
3	Na_2SiF_6	2.0	0.38	6.8	9.5	29.6	44.3	0.37	7.4	10.0	29.9	49.6
4	$CaCl_2$	1.0	0.31	6.0	8.2	16.8	33.4	0.35	7.1	10.9	24.5	45.3
5	$CaCl_2$	2.0	0.32	6.7	8.6	22	33.6	0.36	7.1	10.0	25.1	45.8
6	$FeCl_3$	1.0	0.34	6.9	8.6	19	33.8	0.34	7.2	10.2	26.4	45.6
7	$FeCl_3$	2.0	0.35	6.7	8.9	19.2	35.7	0.36	8.1	10.7	27.2	46.1
8	Na_2CO_3	1.0	0.38	6.0	10.8	20.8	37.5	0.34	7.7	10.2	29.2	46.2
9	Na_2CO_3	2.0	0.4	6.4	10.1	21.2	38.4	0.37	7.7	10.6	31.4	47.4

2.1.3 INTEGRATED METHODS OF ACTIVATION OF LSC

The activating effect of the sulfate additive and the silicon fluoride activator upon their introduction into the LSC is based on various mechanisms of their action. It was of interest to study the possibility of achieving an increased activating effect with the joint introduction of both activators in the LSC.

The main studies were performed using mathematical planning of experiments. For this, a three-level five-factor plan Ha_5 was implemented [40]. The experiment-planning conditions are given in Table 2.5.

In the course of research a cement-sand mortar 1:3 used in which the water-cement ratio (W/C) was determined to achieve a mortar cone spreading on the shaking table of at least 106 mm. The compressive and bending strength of the sample was tested at 7 and 28 days.

After processing and statistical analysis of the experimental data, mathematical models of normal consistency, W/C, and compressive strength of standard mortars based on the studied binders in the form of polynomial regression equations were obtained and are given in Table 2.6.

An analysis of the obtained models (Figure 2.2) indicates that, as expected, the most significant decrease in normal consistency is observed with the introduction of superplasticizer type naphthalene formaldehyde type SP-1 in an amount of 1% by weight of the binder. The normal consistency of such cement is reduced 20% to 22%, that is, a decrease in its water demand is 15% to 20%. As expected, an increase in the specific surface of the low-clinker cement has a significant effect on the normal consistency cement test, namely, when its value increases from $S_{sp} = 400$ to 450 m²/kg to $S_{sp} = 500$ to 550 m²/kg. In this case, the normal consistency of the cement paste is increased by 20% to 23%.

An increase in the content of FG does not significantly affect the normal consistency of the cement paste, but a certain increase in it causes an increase in the content of sodium silicon fluoride additive. It is associated with an increase in the reactivity of slag particles with the addition of fluorite additives [20].

Analyzing equations (2.2) (Table 2.6) and graphical dependencies shown in Figure 2.2 indicates an increase in the clinker content does not significantly affect W/C of the mortars prepared at the LSC. A significant effect on it is the specific surface area of the binder. With an increase in the specific surface 420 to

TABLE 2.5
Experiment Planning Conditions

No.	Factors		Levels of Variation			Range of Variation
	Natural	Coded	−1	0	+1	
1	Clinker content (%)	x_1	5	12	19	7
2	Phosphogypsum content in terms of SO (%)	x_2	3.03	4.5	5.97	1.47
3	Binder specific surface (m²/kg)	x_3	350	450	550	100
4	Hardening activator content (Na_2SiF_6) (%)	x_4	0	2	4	2
5	Superplasticizer content SP-1 (%)	x_5	0	0.5	1	0.5

TABLE 2.6
Experimental-Statistical Models of Cement Paste Normal Consistency, Water-Cement Ratio, and Mortar Strength at LSC

Parameters	Statistical Models	
Normal consistency of cement paste (%)	$NC = 24.8 + 1.1x_1 + 0.2x_2 + 2.4x_3 + 1.3x_4 - 2.4x_5$ $- 0.7x_1^2 + 2x_3^2 - 0.5x_4^2 - 0.2x_5^2 + 0.2x_1x_2 + 1.1x_1x_3$ $- 0.1x_1x_4 + 0.4x_1x_5 + 0.4x_2x_3 + 0.6x_2x_4 + 1x_2x_5$ $+ 0.013x_3x_4 + 0.2x_3x_5 + 0.3x_4x_5$	(2.1)
W/C	$W/C = 0.36 + 0.01x_3 + 0.01x_4 - 0.03x_5$ $- 0.01x_2^2 + 0.02x_3^2 - 0.02x_4^2 + 0.01x_1x_2$	(2.2)
Compressive strength at 7 days (MPa)	$f_c^7 = 19.8 + 1.8x_1 - 2.2x_2 + 2.9x_3 - 0.1x_4 + 0.9x_5$ $- 3.1x_1^2 - 1.4x_2^2 - 1.1x_3^2 - 2.3x_4^2 + 3.5x_5^2 - 0.7x_1x_2$ $- 0.4x_1x_3 - 1.9x_1x_4 - 0.3x_1x_5 + 0.3x_2x_3 - 0.8x_2x_4$ $- 0.5x_2x_5 + 0.7x_3x_4 - 0.4x_3x_5 + 1.9x_4x_5$	(2.3)
Compressive strength at 28 days (MPa)	$f_c^{28} = 44 + 5.73x_1 - 0.7x_2 + 1.49x_3 + 0.22x_4 + 1.19x_5$ $- 6.32x_1^2 - 2.42x_2^2 - 3.22x_3^2 - 4.17x_4^2 + 1.33x_5^2$ $- 0.65x_1x_2 - 1.53x_1x_3 - 2.69x_1x_4$ $- 0.55x_1x_5 + 1.28x_2x_3 + 0.4x_2x_4 - 3.65x_2x_5$ $- 0.86x_3x_4 + 0.2x_3x_5 + 2.91x_4x_5$	(2.4)

450 m²/kg, the water demand of the binder decreases, which is associated with an increase in the plasticity of mortars with an increase in the dispersion of binder particles; however, its excessive increase leads to a sharp increase in W/C. Also, a slight increase in water demand causes an increase in the content of sodium silicon fluoride.

The influence of these two factors is extreme. An increase in the content of FG to 7.5% (4.5% in terms of SO_3) in the total binder mass leads to a slight increase in W/C. This increase is associated with the formation of an increased amount of ettringite as a result of the interaction of the aluminate component of cement clinker with the sulfate component of FG. However, when the aluminate component is completely spent on its formation, a further increase in the content of FG leads to a slight decrease in W/C (Figure 2.3).

The most significant effect of reducing W/C without changing the consistency of the mortar is achieved with the introduction of a superplasticizer. The superplasticizer SP-1 reduces the W/C of the mortar on the LSC from 0.39 to 0.33 and, accordingly, increases the strength of the samples. The effect of superplasticizer on W/C, as follows from the analysis of the obtained model, is almost linear.

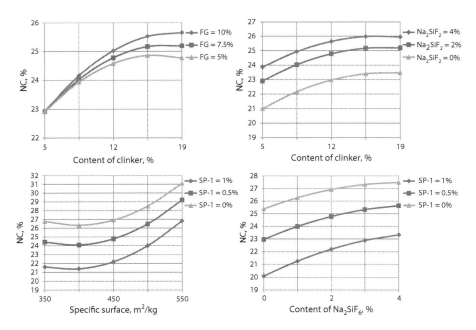

FIGURE 2.2 The influence of technological factors on the normal consistency of the cement test of LSC.

Reducing the water demand of binders due to the introduction of superplasticizer additives and increasing specific surface area can significantly increase their strength. Strength at 7 days also is positively affected by an increase in the content of the Na_2SiF_6—hardening activator additive, but up to a value of no more than 2% of the binder mass (Figure 2.4). This result is due to the protonization of water molecules adsorbed on the electron-accepting centers on the surface of the slag particles by replacing the OH group with an F^- ion. At the same time, the activity of the binder increases and acid-base reactions, specifically hydration reactions, are activated [7]. A further increase in the content of this additive leads to a decrease in strength.

An analysis of Figure 2.5 shows that a simultaneous increase in the content of clinker, FG, and sodium silicon fluoride in the total mass of the binder, with an increase in its specific surface, has a positive effect on the strength of LSC. This result is because in the presence of Ca^{2+} and SO_4^{2-} ions and a high dispersion of particles, hydration of the vitreous phase of granular blast furnace slag is accelerated, resulting in the formation of low-basic calcium hydrosilicates.

An excessive increase in the value of the specific surface of the binder, as well as the consumption of the addition of sodium silicon fluoride, which is due to a sharp increase in W/C, negatively affects the strength of LSC. Studies have established that the greatest strength of the samples is observed at a specific surface area of 450 m²/kg, while the optimal consumption of FG and sodium silicon fluoride in the binder is 7.5% (4.5% in terms of SO_3) and 2%, respectively.

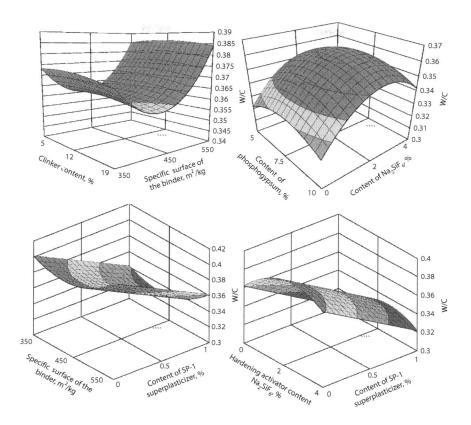

FIGURE 2.3 The influence of technological factors on W/C of mortars based on LSC.

From the results of previous studies, it follows that the introduction of substances, mainly ionic due to the chemical nature of the bond, as well as substances with oxidizing properties, promotes the activation of hydration and increases the strength of LSC. However, the disadvantage of the previously described types of activation of LSC is the low pH of hardening binders and, as a result, the failure of passivation of reinforcing steel.

An effective way to activate slag cement binders, while increasing pH, is the introduction of caustic alkalis. However, the main disadvantage of this method is binding system sets too quickly, which leads to a sharp decrease in the shelf life of concrete and mortars made with such binders.

Of interest was the possibility of enhancing the effect of sulfate-fluoride activation, when using the addition of sodium silicon fluoride (Na_2SiF_6), by the additional introduction of $Ca(OH)_2$ in the LSC. From the general chemical positions in such a system, a chemical reaction should occur:

$$Na_2SiF_6 + 3Ca(OH)_2 = 3CaF_2 + 2NaOH + Si(OH)_4.$$

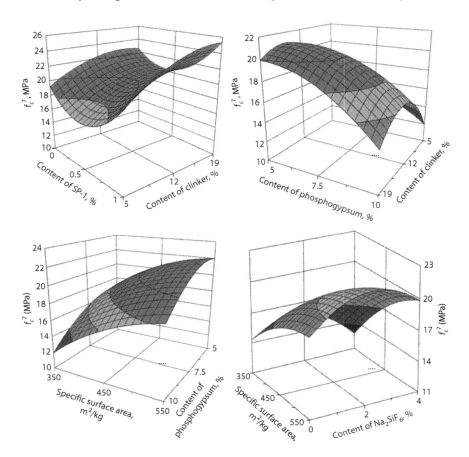

FIGURE 2.4 The influence of technological factors on the strength of mortars based on LSC at 7 days.

In the hardening cement mortar, along with calcium fluoride (CaF_2), a caustic alkali (NaOH) forms, this provides additional alkaline activation. The results of a study of the effectiveness of sulfate-fluoride-alkaline activation (SFA) are given in Table 2.7.

All studies were carried out on the same binder of the following composition: Portland cement clinker, 12%; blast furnace granulated slag, 88%; and phosphogypsum dihydrate, 7.5% (4.5% in terms of SO_3).

Analysis of the obtained results indicated the introduction of a complex additive in the form of a mixture of sodium silicon fluoride (Na_2SiF_6) and lime into the composition of LSC can significantly improve the properties of the binder, namely, the setting time is significantly reduced (from 3 hours 10 minutes to 1 hour 15 minutes) and pH of the medium increased (from 9–10 to 12–13); the enhancement of LSC activity at 7 and 28 days almost doubled. These characteristics significantly expand the scope of application of this cement; namely, it can be used for the manufacture of reinforced concrete structures, since the pH of the medium will ensure the passivation of reinforcing steel in concrete.

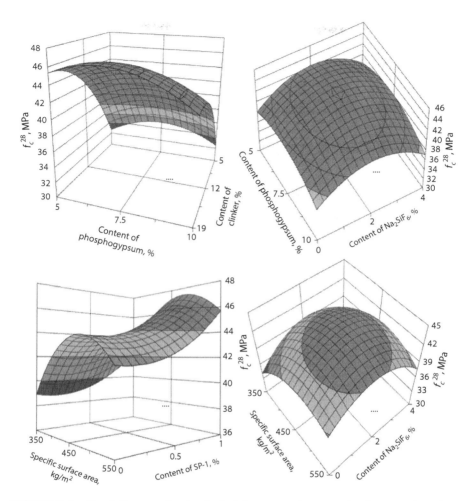

FIGURE 2.5 The influence of technological factors on the strength of mortars based on LSC at 28 days.

2.1.4 Features of Hydration and Structure Formation of the LSC

The features of the structure formation of cement pastes based on LSC were studied using plastograms and velocity curves of the passage of ultrasonic waves through hardening pastes (Figures 2.6 and 2.7).

An analysis of the plastogram shows that the LSC is characterized by a slightly increased period of coagulation structure formation, due to the fact that after a few minutes the surface of the slag particles is covered with hard-to-permeable layer of $Al(OH)_3$, $Si(OH)_4$, and hydro- and aluminosilicates [69]. In cement pastes at LSC, superplasticizer and activator promote growth of more intensive plastic strength on another section of plastogram.

TABLE 2.7
Test Results of Sulfate Fluoride Alkaline Activation LSC

No.	Kind of Binder	W/C	SC (mm)	$f_{c,tb}^{7}$ (MPa)	f_{c}^{7} (MPa)	$f_{c,tb}^{28}$ (MPa)	f_{c}^{28} (MPa)	pH
1	LSC	0.4	130	2.0	17	5.6	29.4	9.5
2	LSC	0.5	215	2.6	13.3	3.2	24.1	9.3
3	LSC + Lime (3%)	0.4	120	3.6	19.9	8.2	36.3	12.5
4	LSC + Lime (3%)	0.5	205	2.9	16.3	5.9	30.8	12.8
5	LSC + Lime (7%)	0.4	200	3.0	17.8	6.9	31.4	12.7
6	LSC + Lime (7%)	0.5	120	2.8	15.5	6.1	30.2	12.9
7	LSC + Na$_2$SiF$_6$ (2%)	0.4	120	4.4	24.7	6.7	38.3	10.1
8	LSC + Na$_2$SiF$_6$ (2%)	0.5	205	3.9	21.3	6.7	30.5	10.2
9	LSC + Lime (3%) + Na$_2$SiF$_6$ (2%)	0.4	115	6.7	28.9	7.2	46.7	12.6
10	LSC + Lime (3%) + Na$_2$SiF$_6$ (2%)	0.5	200	6.4	24.3	7.5	39.2	12.5
11	LSC + Lime (7%) + Na$_2$SiF$_6$ (2%)	0.4	115	7.1	24.9	8.5	44.9	12.8
12	LSC + Lime (7%) + Na$_2$SiF$_6$ (2%)	0.5	195	5.4	23.3	7.2	35.4	12.5

Note: SC, spreading cone, $f_{c,tb}^{7}, f_{c}^{7}, f_{c,tb}^{28}, f_{c}^{28}$, accordingly, bending and compression strength at 7 and 28 days.

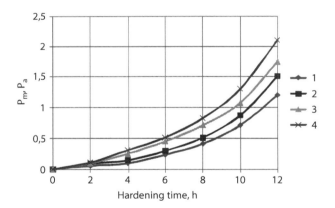

FIGURE 2.6 Kinetics of the plastic strength of cement pastes on LSC: (1) LSC without additives, (2) LSC + SP-1 (0.6%), (3) LSC + Na$_2$SiF$_6$ (2%), and (4) LSC + Na$_2$SiF$_6$ (2%) + quicklime (3%).

In the plastogram, the second section corresponds to the period of strengthening of the coagulation structure, as well as the beginning of the formation of the crystallization structure. The introduction of hardening activators (Na$_2$SiF$_6$ and mixtures of Na$_2$SiF$_6$ + quicklime) in this area leads to an increase in plastic strength. After 10 to 12 hours of hardening, the plastic strength of the pastes obtained by activated LSC

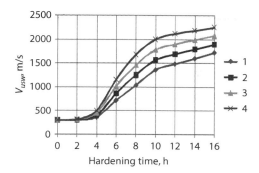

FIGURE 2.7 Changing the speed of ultrasound in cement pastes made at LSC: (1) LSC without additives, (2) LSC + SP-1 (0.6%), (3) LSC + Na_2SiF_6 (2%), (4) LSC + Na_2SiF_6 (2%) + Quicklime (3%).

significantly increases and reaches 2.1 Pa. To achieve such a value of plastic strength for inactivated pastes requires a much more time.

The kinetic in the gain of plastic strength (Figure 2.6) was compared with the kinetic of the velocity of ultrasonic waves passing through hardening pastes. Using the ultrasonic impulsive method [79], the propagation velocity of the leading edge of the ultrasonic wave (V_{usw}, m/s) was determined. To measure the propagation time of ultrasound, the sensors were installed on opposite sides of the material. As can be seen from Figure 2.7, the curves of the dependence of the speed of passage of ultrasonic waves through cement pastes in time contain areas that also characterize the stages of structure formation: the initial induction period, the period of growth and adhesion of gel-like (calcium hydrosilicates) and crystalline neoplasms (hydrosulfoaluminates), and the period of the final structure strengthening with subsequent recrystallization of neoplasms.

The smallest strength, and, accordingly, the speed of ultrasound passing through a hardening binder, is typical for LSC, which does not contain additives. The use of superplasticizer, as well as additives of hardening activators, leads to an increase in the speed of passage of ultrasonic waves and the strength of the binder due to a decrease in its water demand and an increase in hardening speed. The maximum speed of ultrasonic waves is characteristic for a binder containing a complex activator additive, consisting of sodium silicon fluoride and quicklime.

The mechanical properties of cement stone and concrete significantly depend not only on the chemical composition of hydrated cement, but also on the composition and structure of hydration products.

Peculiarities of the phase composition and microstructure of cement stone based on LSC were studied using X-ray diffraction analysis and electron microscopy.

A characteristic feature of binders that contain a large amount of blast furnace granulated slag is the presence of a sufficiently large amount of active amorphous glassy phase.

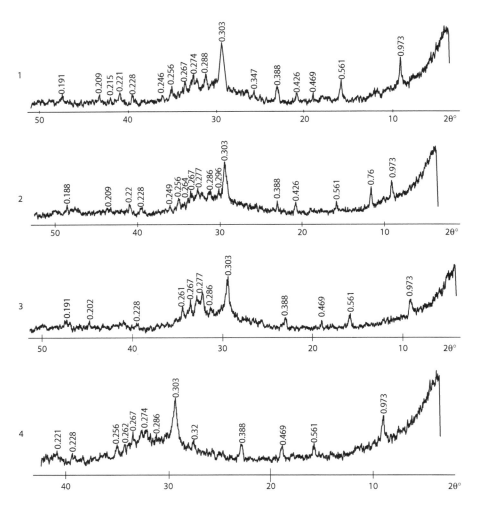

FIGURE 2.8 X-ray pictures of hardened binder: (1) C-1, hydrated at 28 days, (2) CC (C, 50%; Ash, 50%; S_{sp} = 460 m²/kg), (3) CC (C, 50%; As, 50%; S_{sp} = 460 m²/kg), and (4) CC (C, 50%; Ash, 37%; Sl, 12.5%; Sika VC 225, 0.7%; S_{sp} = 460 m²/kg).

The analysis of X-ray diffraction patterns (Figure 2.8) indicates that the hydration processes of LSC are actively taking place with the formation of a high sulfate form of calcium hydrosulfoaluminate $3CaO \cdot Al_2O_3 \cdot 3CaSO_4 \cdot 32H_2O$ (d/n = 0.973; 0.561; 0.469; 0.388; 0.2564; 0.221; 0.2154 nm) (ettringite) [70].

The presence of lines of $CaSO_4 \cdot 2H_2O$ indicates that on day 7 of hardening, not all gypsum chemically bonded to hydrosulfoaluminates. But the intensity of the gypsum lines (d/n = 0.76; 0.427; 0.379; 0.3059 nm) significantly decreased at 28 days.

The results obtained generally agree with the data of [76] according to which, with an increase in the hardening time, the content of two-water gypsum and portlandite decreases while the content of ettringite increases.

Comparison of diffractograms shows that the overall picture of the mineralogical composition with the introduction of superplasticizer practically does not change.

A significant difference between the X-ray diffraction pattern is the presence in the cement stone of appreciable amounts of portlandite (d/n = 4.93; 2.63; 1.93; 1.79 nm), which, after 7 days of hardening, has not yet completely bound into hydrosilicates. In addition, on this X-ray there are lines of ettringite, hydrosilicates, and gypsum.

Electron microscopic studies were carried out by the replica method on an electron microscope REMMA-101-02 of samples by spraying copper or graphite onto the cleaved surface.

A comparison of electron microscopic images of hardened LSC samples at 7 and 28 days of hardening indicates an increase in the number of needle-shaped and prismatic crystals of ettringite and a decrease in the content of lamellar crystals of two-water gypsum with an increase in the duration of hardening of the binder (Figure 2.9). At the same time, there is a disappearance of signs of the presence of portlandite, which are clearly visible in samples that hardened for 7 days.

The use of Sika VC 225 superplasticizer in a certain way affected the morphology of solidified stone components at 7 and 28 days. Prismatic structures are clearly visible, which can be attributed to gypsum, as well as hydrosulfoaluminate. Fibrous structures indicate the presence of low-basic calcium hydrosilicates, which, as mentioned previously, according to the results of X-ray phase analysis, indicates an increase in the proportion of tobermorite calcium hydrosilicates in the products of hydration of LSC.

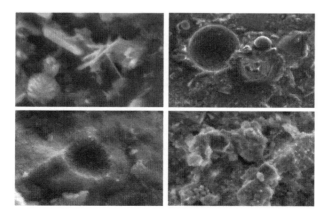

FIGURE 2.9 Microphotographs of various areas of hardened stone LSC.

2.2 NORMAL-WEIGHT CONCRETES BASED ON ACTIVATED LSC

According to the results of previous studies, it was found that combined SFA activation is the most effective of all considered activation types. The effect of the activated LSC and various plasticizing additives content on the water demand of concrete mixtures and the strength of concrete was studied. A three-level two-factor plan was implemented [40]. Experimental planning conditions are listed in Table 2.8. All concrete mixes were made based on the same binder of the following composition: clinker, 12%; slag, 8%, FG, 7.5% (SO_3, 4.5%). To ensure SFA activation, in addition to the binder, lime was introduced in the amount of 3% by the binder weight, and sodium silicon fluoride (Na_2SiF_6) in the amount of 2%. The strength of the binder with a specific surface of 453 m²/kg was 45 MPa. Plasticizing additives were technical lingosulfonate (LST), naphthalene-formaldehyde SP-1 type superplasticizer, and polycarboxylic superplasticizer Sika VC 225.

As aggregates for concrete, granite crushed stone with a maximum grain size of 20 mm and sand with a fineness modulus of 1.9 were used.

The slump cone of concrete at all points in the plan was 5 to 9 cm. After processing and statistical analysis of experimental data, mathematical models of the concrete mix water demand and standard concrete cubes compressive strength, in a form of polynomial regression equations, are obtained. Also, on the basis of the obtained experimental data, it was possible to construct additional mathematical models of concrete compressive strength at 28 days, in which the W/C (X_1') and consumption of plasticizing additive (X_2') were chosen as variable factors. The results of processing and statistical analysis of the experimental data are shown in Table 2.9.

Graphs illustrating the influence of technological factors on the concrete mix water demand, concrete compressive strength after heat treatment, and 28 days of normal curing are shown in Figures 2.10 through 2.13.

Studies have shown that the most significant influence on the concrete mixture water demand is the consumption of plasticizer (Figure 2.10). When the superplasticizer content is increased to 3 kg/m³, the concrete mix water demand is reduced by 20% when using the SP-1 plasticizing additive, and by 35% when using Sika VC 225 and Sika VC 225+LST plasticizing additives. It also was found that with a superplasticizer content of 1.5 kg/m³, the highest water-reducing effect was observed when

TABLE 2.8
Experimental Planning Conditions

No.	Factors Natural	Coded	Variation Levels −1	0	+1
1	Binder content (kg/m³)	x_1	300	400	500
2	Plasticizing admixtures content (kg/m³): SP-1, Sika VC 225, Sika VC 225+LST(1:1)	x_2	0	1.5	3

TABLE 2.9
Experimental-Statistical Models of Concrete Strength at Activated LSC

Type Plasticizer	Statistical Models	
Concrete Water Demand		
SP-1	$W = 177.3 + 6.168x_1 - 13.169x_2 + 0.936x_1^2$ $- 0.064x_2^2 - 3.5x_1x_2$	(2.5)
Sika VC 225	$W = 145.3 + 1.667x_1 - 32.507x_2 + 5.894x_1^2$ $+ 11.894x_2^2 - 7.0x_1x_2$	(2.6)
Sika VC 225+LST	$B = 157.1 + 3.334x_1 - 28.172x_2 + 3.285x_1^2$ $+ 7.785x_2^2 - 10.75x_1x_2$	(2.7)
Compressive Strength at 7 days		
SP-1	$f_c^7 = 16.118 + 6.985x_1 + 2.217x_2 + 3.014x_1^2$ $+ 1.114x_2^2 + 1.2x_1x_2$	(2.8)
Sika VC 225	$f_c^7 = 17.656 + 7.552x_1 + 6.651x_2 + 6.969x_1^2$ $+ 1.369x_2^2 + 1.2x_1x_2$	(2.9)
Sika VC 225+LST	$f_c^7 = 25.372 + 5.251x_1 + 3.817x_2 - 3.618x_1^2$ $- 2.118x_2^2 - 0.2x_1x_2;$	(2.10)
Compressive Strength at 28 days		
SP-1	$f_c^{28} = 29.612 + 10.252x_1 + 2.517x_2 - 1.764x_1^2$ $+ 0.236x_2^2 + 0.2x_1x_2;$	(2.11)
	$f_c^{28} = 29.702 - 15.663x_1' + 0.139(x_1')^2 - 0.191(x_2')^2;$	(2.12)
Sika VC 225	$f_c^{28} = 42.659 + 12.036x_1 + 12.736x_2$ $+ 1.275x_1^2 - 4.625x_2^2 + 1.2x_1x_2;$	(2.13)
	$f_c^{28} = 31.953 - 21.844x_1' - 0.016(x_1')^2 - 0.016(x_2')^2;$	(2.14)
Sika VC 225+LST	$f_c^{28} = 38.988 + 9.519x_1 + 6.851x_2 - 5.451x_1^2$ $- 2.351x_2^2 + 1.575x_1x_2;$	(2.15)
	$f_c^{28} = 30.922 - 15.568\ x_1' - 0.02(x_1')^2 - 0.015(x_2')^2$	(2.16)

(Continued)

TABLE 2.9 (*Continued*)
Experimental-Statistical Models of Concrete Strength at Activated LSC

Type Plasticizer	Statistical Models	
Compressive Strength After Heat Treatment[a]		
SP-1	$f_c^{ht} = 27.31 + 10.035x_1 + 2.851x_2 - 3.313x_1^2$ $+ 0.637x_2^2 + 1.625x_1x_2;$	(2.17)
	$f_{cm}^{ht} = 27.1 - 14.1x_1' + 0.12x_2' - 1.9(x_1')^2 - 0.2\,x_1'x_2'$	(2.18)
Sika VC 225	$f_c^{ht} = 36.076 + 7.451x_1 + 7.968x_2 - 4.123x_1^2$ $- 2.473x_2^2 - 2.45x_1x_2;$	(2.19)
	$f_{cm}^{ht} = 26.9 - 12.5x_1' + 0.1(x_1')^2 + 0.01(x_2')^2$	(2.20)
Sika VC 225+LST	$f_c^{ht} = 29.921 + 6.118x_1 + 3.934x_2 - 5.789x_1^2$ $+ 0.761x_2^2 + 0.8x_1x_2;$	(2.21)
	$f_{cm}^{ht} = 29.7 - 11.6x_1' - 0.8x_2' - 3.5(x_1')^2$ $- 1.4(x_2')^2 - 0.23x_1'x_2'$	(2.22)
Compressive Strength After Heat Treatment and 28 Days of Normal Hardening[a]		
SP-1	$f_c^{ht28} = 33.435 + 14.086x_1 + 3.701x_2$ $- 2.92x_1^2 + 3.53x_2^2 + 2.05x_1x_2;$	(2.23)
	$f_{cm}^{ht28} = 33.6 - 19.2x_1' + 0.95x_2' - 0.8(x_1')^2$ $- 0.24(x_2')^2 - 1.05x_1'x_2'$	(2.24)
Sika VC 225	$f_c^{ht28} = 54.041 + 14.236x_1 + 15.636x_2$ $- 2.576x_1^2 - 5.736x_2^2 + 2.55x_1x_2;$	(2.25)
	$f_{cm}^{ht28} = 38.5 - 31.9x_1' + 0.1x_2' + 8.3(x_1')^2$ $+ 1.89(x_2')^2 - 0.1x_1'x_2'$	(2.26)
Sika VC 225+LST	$f_c^{ht28} = 45.727 + 10.969x_1 + 6.168x_2 - 7.346x_1x_2$ $- 1.146x_2^2 - 2.4x_1x_2;$	(2.27)
	$f_{cm}^{ht28} = 41.1 - 15.6x_{1'} + 0.1x_{2'} - 6.7(x_{1'})2$ $- 0.02(x_{2'})2 - 0.1x_{1'}x_{2'}$	(2.28)

[a] Heat treatment of concrete was carried out by steaming at 80°C. The rate of the temperature rise and cooling was 30°C per hour. The isothermal aging duration is 6 hours.

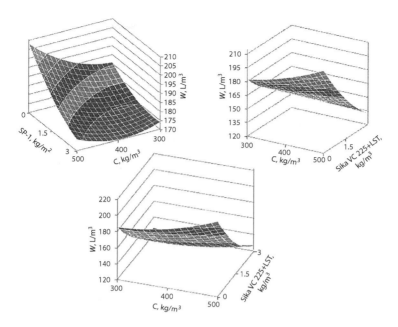

FIGURE 2.10 Influence of technological factors on the water demand of concrete made based on activated LSC.

using Sika VC 225 additive, less when using Sika VC 225 and LST additives in quantities of 1:1, and the least when using SP-1 additive.

LSC-based concrete strength at 7 and 28 days increases most significantly with decreasing W/C. Increasing the content of Sika VC 225 to 3 kg/m^3 doubles concrete strength. The obtained strength values are 40 MPa at 7 days and more than 60 MPa at 28 days. Using Sika VC 225+LST achieves nearly 30 MPa at 7 days and about 45 MPa at 28 days. For SP-1 the strength at 7 and 28 days corresponds to 30 and 40 MPa, respectively (Figure 2.11).

The relatively high strength of concrete, based on LSC, can be explained by high reactivity of the binder that has very fine milling, yielding a higher activation effect of cement and sulphate compounds of the binder for the blast furnace slag. Achieving low W/C by using plasticizing admixtures promotes cement hydration in "cramped conditions" that yields a more rapid formation of oversaturated solution, at which new hydrate substances are formed rapidly [71].

The more significant effect of using fine milled LSC is observed at heat treatment (Figures 2.12 and 2.13). A characteristic feature of concrete based on LSC is intensive strength growth after steaming. The specimen's strength at 4 hours after steaming is 45 MPa and at 28 days steamed specimens, made using Sika VC 225 plasticizing admixture, achieve 80 MPa. When Sika VC 225+LST is added to the

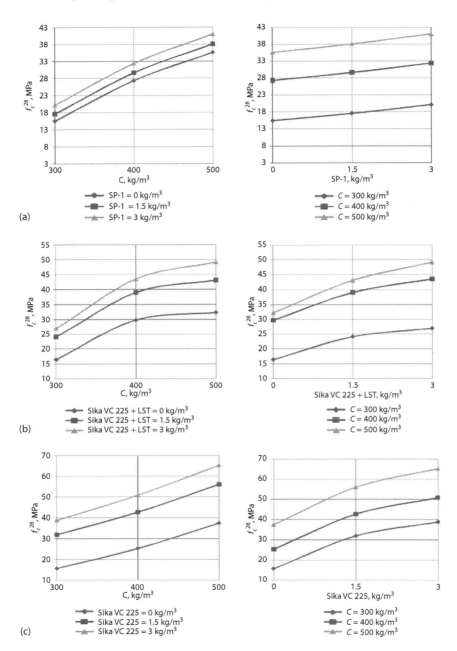

FIGURE 2.11 Influence of technological parameters on LSC based on concrete strength at 28 days: (a) SP-1 superplasticizer, (b) Sika VC 225+mechnical lignosulfonate (LST) (1:1), and (c) Sika VC 225 superplasticizer.

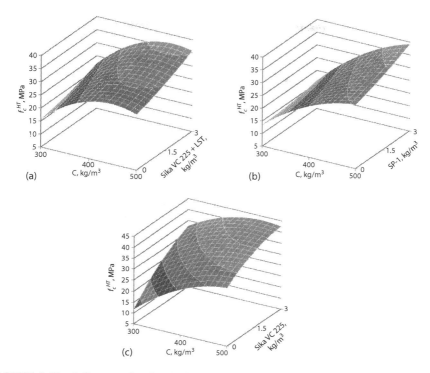

FIGURE 2.12 Influence of technological parameters on LSC based on concrete strength after heat treatment: (a) Sika VC 225+LST (1:1) superplasticizer, (b) SP-1 superplasticizer, and (c) Sika VC 225 superplasticizer.

concrete, the strength values are 35 MPa and 50 MPa after 4 hours of steaming and at 28 days, respectively, whereas when adding SP-1, the strength values are 40 MPa and 50 MPa, respectively (Figures 2.12 and 2.13).

Mathematical models of concrete compressive strength at 28 days, in which the W/C (x_1') and the plasticizing additive contents (x_2') were selected as variable factors (Table 2.9), allow to calculate the compositions of normal weight concrete made at LSC with a given strength and workability. The methodology includes the following steps:

1. Selecting the type and content of plasticizer and the concrete mix workability.
2. Using the experimentally obtained dependences of the water demand (W), for a given concrete mix workability value, the plasticizer type and content water consumption is determined.
3. To determine the W/C of the concrete mixture, mathematical models of concrete compressive strength are used, in which the W/C (x_1') and the consumption of the plasticizing additive (x_2') were selected as variable factors,

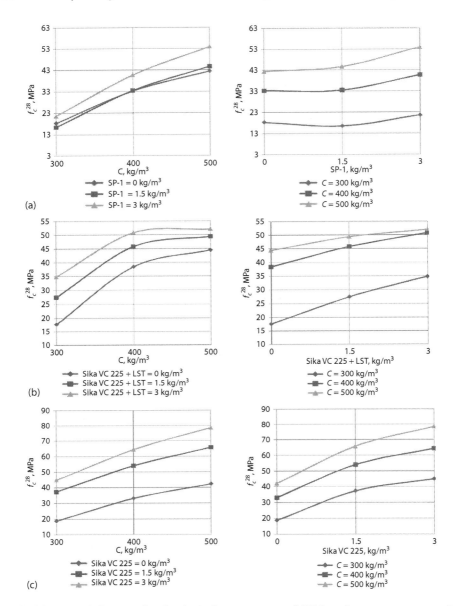

FIGURE 2.13 Influence of technological parameters on LSC based on concrete strength after heat-humid processing: (a) SP-1 superplasticizer, (b) Sika VC 225+LST (1:1) superplasticizer, and (c) Sika VC 225 superplasticizer.

having previously expressing the content of the plasticizing additive into a coded form by the formula:

$$x_2' = \frac{(SP - 0.3)}{0.3}. \tag{2.29}$$

4. The obtained value of the W/C, x_1', is transformed into natural appearance, given that:

$$x_1' = \frac{(W/C - 0.45)}{0.2}. \tag{2.30}$$

5. Knowing the water consumption and the W/C, find the consumption of cement:

$$C = W \frac{1}{W/C} \tag{2.31}$$

6. Consumption of crushed stone (CS) and Sand (Sd) are found by known equations:

$$CS = \frac{1000}{\alpha \dfrac{P_{cs}}{\rho_{b.cs}} + \dfrac{1}{\rho_{cs}}}; \tag{2.32}$$

$$Sd = \left(1000 - \left(\frac{C}{\rho_C} + W + \frac{CS}{\rho_{CS}}\right)\right)\rho_{Sd}, \tag{2.33}$$

where α is the grain spread coefficient, ρ_{CS} is the density of crushed stone, $\rho_{b.sc}$ is the bulk density of crushed stone, ρ_{sd} is the density of sand, and P_{CS} is the intergranular emptiness of crushed stone.

2.2.1 CALCULATION EXAMPLE

The composition of normal weight concrete was calculated based on LSC with the required compressive strength of 35 MPa after heat treatment and 70 MPa after 28 days of further normal curing. The slump cone of the concrete mixture is equal to 15 cm. As a plasticizing additive, 3 kg/m³ of superplasticizer Sika VC is used.

1. According to (b) in Figure 2.14, for the given concrete mix workability ($S = 15$ cm), plasticizer type (Sika VC 225) and content (3 kg/m³), the water demand is 135 L/m³.
2. Transform the plasticizing additive content into coded form:

$$x_2' = \frac{(3 - 1.5)}{1.5} = 1$$

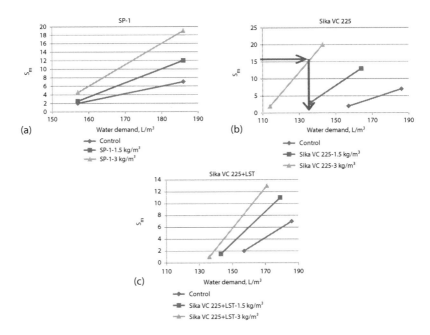

FIGURE 2.14 Dependences of cone slump for LSC based on concrete vs. water demand per cubic meter: (a) SP-1 superplasticizer, (b) Sika VC 225 superplasticizer, and (c) Sika VC 225+LST (1:1) superplasticizer.

3. From equations (2.20) and (2.26) (Table 2.9), W/C (X_1') is determined, which will provide the required strength of concrete.
After passing the heat treatment ($f_c^{ht} \geq 40$ MPa):

$$35 = 26.9 - 12.5x_1' + 0.1(x_1')^2 + 0.01(1)^2$$

$$-0.1(x_1')^2 - 12.5x_1' - 8.1 = 0.$$

Solving the obtained quadratic equation gives: $x_1' = -0.64$.
After the heat treatment and 28 days of further normal curing ($f_c^{ht28} \geq 70$ MPa):

$$70 = 38.5 - 31.9x_1' + 0.1 \cdot 1 + 8.3(x_1')^2 + 1.89(1)^2$$

$$0.1x_1' \cdot 1 - 8.3(x_1')^2 + 31.8\, x_1' + 33.29 = 0.$$

Solving the obtained quadratic equation gives: $x_1' = -0.86$.

4. Transform the obtained values of the W/C in natural appearance:
After heat treatment:

$$\frac{W}{C} = x_1' \times 0.2 + 0.45 = -0.64 \times 0.2 + 0.45 = 0.32;$$

after heat treatment and 28 days of normal curing:

$$\frac{W}{C} = x_1' \times 0.2 + 0.45 = -0.86 \times 0.2 + 0.45 = 0.28.$$

To ensure the required concrete strength characteristics, the minimum value of the W/C, which is used for further calculations, is accepted.

5. Determine the required consumption of cement:

$$C = \frac{135}{0.28} = 482 \, \text{kg/m}^3.$$

6. The aggregates contents are found at the grain spread coefficient $\alpha = 1.46$, the density and bulk density of crushed stone $\rho_{CS} = 2.85$ and $\rho_{b.CS} = 1.65$ kg/l density of sand $\rho_{sd} = 2.65$ kg/l, and intergranular emptiness of crushed stone $P_{CS} = 0.42$ [31].

$$CS = \frac{1000}{1.46 \frac{0.42}{1.65} + \frac{1}{2.85}} = 1388 \, \text{kg/m}^3;$$

$$S = \left(1000 - \left(\frac{482}{3.1} + 135 + \frac{1388}{2.85}\right)\right) \times 2.65 = 590 \, \text{kg/m}^3.$$

The calculated concrete mix composition: cement, 482 kg/m³; water, 135 L/m³; crushed stone, 1388 kg/m³; sand, 590 kg/m³; and Superplasticizer Sika VC 225, 3 kg/m³.

Thus, studies have shown the possibility of obtaining concretes, including increased strength, using activated LSC as a binder, in which clinker contains less than 20%. Concrete strength after 28 days of curing in normal conditions is above 50 MPa. The higher effect of applying fine-milled LSC is observed using heat curing. Using polycaroxilate type hyperplasticizers Sika VC allows the use of LSC for producing concrete that after heat curing at 80°C has a compressive strength of about 80 MPa.

2.2.2 Foam Concrete

Foam concrete is a composite material obtained by mixing aqueous solution of binders and foam. The quality of the foam concrete depends directly on the raw materials used and the structure that is formed directly during its preparation and during curing [72]. The foam concrete quality is significantly influenced by the fineness of grinding of the binder used [73]. Increasing the cement fineness of the grinding should contribute to the formation of a more homogeneous interstitial partitions and, therefore, a better structure of the foam concrete [73].

A comparative analysis of the study results on this problem suggests that the physico-mechanical properties of foam concrete based on low-clinker cements, under certain conditions, are no worse than properties of the material based on Portland cement [74–79].

The aim of our research was to determine the possibility of obtaining non-autoclaved foam concrete using LSC, with a clinker content of less than 20%.

The LSC with clinker content of 12% was used as the raw material for the experimental studies. The binder strength with a specific surface area of 450 m²/kg was 43 MPa. The plasticizing additives were a superplasticizer of naphthalene formaldehyde type SP-1, as well as a superplasticizer of polyacrylate type Dynamon SP_3. The foaming agent was used to obtain a stable foam. The mineral filler used was ground sand with a specific surface area of 410 m²/kg.

The main studies were performed using mathematical planning of the experiment. For this purpose, a three-level three-factor plan B_3 was implemented [40]. The experimental planning conditions are given in Table 2.10.

After conducting, processing, and statistical analysis of the experimental data, mathematical models of average density and compressive strength of standard concrete cubes that underwent heat treatment and cubes after 28 days of normal curing were obtained in the form of polynomial regression equations (Table 2.11). Based on the obtained experimental-statistical models the graphical dependencies (Figures 2.15 to 2.17) are obtained.

Analysis of graphical dependences in Figure 2.15 indicates that the most significant factors affecting the foam concrete average density are the contents of the binder and the foaming agent. The graph shows that increasing the amount of binder leads to a sharp increase in the average density, which is associated with an increase in the amount of solid phase. However, this effect can be offset by an increase in

TABLE 2.10
Experimental Planning Conditions

	Factors		Variation Levels			
No.	Natural	Coded	−1	0	+1	Variation Interval
1	LSC consumption (C) (kg/m³)	x_1	600	700	800	100
2	Filler to cement ratio (F/C) by weight	x_2	0	0.25	0.5	0.25
3	Foaming agent consumption (kg/m³)	x_3	1	1.5	2	0.5

Note: Clinker consumption ranges from 72 to 96 kg/m³.

TABLE 2.11
Experimental-Statistical Models of Water Demand and Foam Concrete Strength Based on LSC

Parameters	Statistical Models	
Foam concrete water demand (L/m³)	$B = 264.84 + 17.5x_1 - 0.07x_2$ $-1.33x_1^2 - 2.33x_2^2 + 1.17x_3^2$	(2.34)
Average density (g/cm³)	$\rho = 0.71 + 0.1155x_1 + 0.0337x_2$ $-0.152x_3 + 0.072x_1^2 + 0.076x_2^2 + 0.076x_3^2$ $+0.014x_1x_3 - 0.025x_1x_2 + 0.026x_2x_3$	(2.35)
Compressive strength after heat treatment (MPa)	$f_c^{ht} = 0.27 + 1.423x_1 - 1.051x_2 - 0.172x_3$ $+1.197x_1^2 + 0.407x_2^2 + 1.172x_3^2 - 0.194x_1x_3$ $+0.609x_1x_2 - 0.376x_2x_3$	(2.36)
Compressive strength at 28 days (MPa)	$f_c^{28} = 1.19 + 0.014x_1 + 0.014x_2 - 0.113x_3$ $-3.506x_1^2 + 0.523x_1x_3 + 0.0154x_1x_2$ $-0.047x_2x_3$	(2.37)

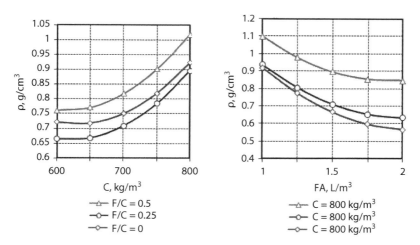

FIGURE 2.15 Influence of technological factors on the average density of foam concrete produced at LSC.

the amount of foaming agent that entails an increase in the volume of foam and a corresponding increase in the hardened concrete porosity. The increasing of the mineral filler amount (milled sand) to F/C = 0.25 leads to a slight decrease in the foam concrete average density, which is associated with an increase in the amount of water required to obtain a concrete mixture of a given mobility, and, accordingly, to increase the porosity of the hardened cement stone and reducing its average density.

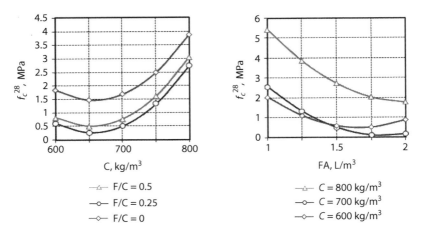

FIGURE 2.16 Influence of technological factors on the foam concrete strength at 28 days produced at LSC.

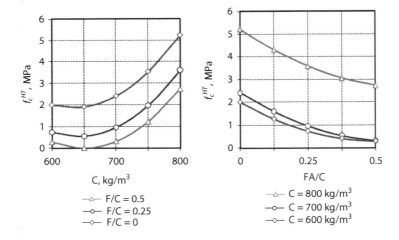

FIGURE 2.17 Influence of technological factors on the foam concrete strength after heat treatment produced at LSC.

The influence of technological factors on the foam concrete strength after passing the heat treatment is like their effect on the strength at 28 days with normal curing (Figures 2.16 and 2.17). However, the foam concrete strength after heat treatment is significantly higher than the strength after 28 days of normal curing, which is caused by the formation at an elevated temperature of an additional amount of calcium hydrosilicates and a dense fine porous structure of cement stone [71].

The second stage of the research was devoted to determining the effect of the superplasticizer type and the amount of foaming agent on the foam concrete properties produced at LSC. For this purpose, several studies were conducted, which also established the optimal parameters for obtaining high quality foam concrete. The results of the studies are given in Table 2.12.

TABLE 2.12
Experimental Results of Foam Concrete Studies Based on the LSC

No.	LSC (kg/m³)	Type of Superplasticizer	Foaming Agent PB-2000, (kg/m³)	W_f (L/m³)	W_m (L/m³)	W_t (L/m³)	ρ_c (kg/m³)	f_c^{ht} (MPa)	CCQ
1	800	SP-1 (0.5%)	2.5	100	150	250	623	1.6	2.57
2	800	SP-1 (0.5%)	2.25	100	150	250	598	1.2	2.01
3	800	SP-1 (0.5%)	2	100	150	250	697	1.8	2.58
4	800	SP-1 (0.5%)	1.8	70	155	225	869	3	3.45
5	800	SP-1 (0.5%)	1.6	70	155	225	867	3.4	3.92
6	800	SP-1 (0.5%)	1.4	70	155	225	816	4.2	5.15
7	800	SP-1 (0.5%)	1	70	155	255	889	4.1	4.61
8	800	Dynamon (0.5%)	1.8	70	155	255	852	3.5	4.11
9	800	Dynamon (0.5%)	1.6	70	155	255	924	3.0	3.25
10	800	Dynamon (0.5%)	1.4	70	155	255	844	2.75	3.26
11	800	Dynamon (0.5%)	1.2	70	155	255	923	3.8	4.12
12	800	Dynamon (0.5%)	1	70	155	255	955	4.9	5.13

Note: W_f is the water consumption for the preparation of foam; W_m is the consumption of water for the preparation of the mortar; W_t is the total water consumption; ρ_c is the average density of samples that have undergone heat treatment; f_c^{ht} is the foam concrete strength after heat treatment; and CCQ is the coefficient of constructive quality.

Analyzing the data obtained, the use of LSC for the preparation of foam concrete is reasonable, which is justified by the sufficiently high values of strength and qualitative porous structure.

The use of the superplasticizer naphthalene-formaldehyde composition can significantly reduce the water demand of the concrete mixture for the manufacture of non-autoclaved foam concrete, resulting in a significant increase in the foam concrete compression strength. The use of a naphthalene-formaldehyde superplasticizer and LSC with a high content of blast-furnace slag improves the structural characteristics of the concrete mix and the hardened foam concrete, increases the uniformity of pore sizes, and provides pronounced geometric ordering, which leads to increased foam concrete strength.

The stability of the pore structure is achieved by the action of the ground granular blast furnace slag and hydrocarbon foaming agent, which reduces the synergy of the foam and thus slows down destructive processes in its structure. In addition, ground granular blast furnace slag promotes moisture retention in the raw mixture during its hardening, which increases the ultimate strength of the foam concrete [80].

Studies have shown the possibility of obtaining high quality foam concrete using LSC with clinker content less than 20%. Using the coefficient of constructive quality (the ratio of the compressive strength to their average density) can set the optimal technological parameters to obtain foam concrete with specified strength and density.

2.2.3 Vibropressed Sawdust Concrete

The purpose of the research was to establish the possibility of obtaining lightweight concrete based on aggregate from woodworking wastes and mineral binder in the form of an LSC.

To achieve this goal, planned experiments were performed with a variation of the sawdust and cement consumptions ratio and the content of mineralizer and hardening accelerator—$CaCl_2$. The experimental planning conditions are given in Table 2.13.

In our studies, sawdust and shavings obtained from softwood lumber were used as aggregates for the preparation of sawdust concrete. The sawdust content (<5 mm) was 80% of the aggregate total weight, the shavings (5–25 mm) 20%, respectively. This particle size distribution, as indicated in [81], provides the formation of the most homogeneous structure of sawdust concrete. LSC with a clinker content of 12% was used as mineral binder.

Compaction of the mixture was carried out by vibration with a load of 0.06 to 0.09 MPa. The formation time of all samples was 20 seconds. To avoid cracking and reduce the elasticity of the wood, sawdust was kept in a mineralizer solution before mixing and formation.

After processing and statistical analysis of experimental data, mathematical models of the W/C, average density, as well as concrete strength at 7 and 28 days were obtained (Table 2.14). Based on the obtained experimental-statistical models, graphical dependencies are given in Figures 2.18 and 2.19.

As evidenced by the obtained experimental data, based on the LSC it is possible to produce structural and thermal insulating sawdust concrete with an average density of 1,450 kg/m³ to 1,800 kg/m³ and a compressive strength of 11 MPa to 18 MPa. The low level of alkaline environment of the LSC has a positive effect on the compatibility of this binder with the wood aggregate and does not cause significant extraction of the wood extractives. The high specific surface area of the LSC leads to an increase in the contact surface of the mortar with the aggregate, which promotes the adhesion of the binder to the sawdust and has a positive effect on the sawdust concrete strength.

The samples of sawdust concrete without the calcium chloride addition are characterized by a slow set of strength, especially in the early periods of hardening. The studies have shown that an increase in $CaCl_2$ content of up to 2% by the binder weight leads to an increase in the sample strength, at 7 days, on average, by 50% to 55%, and at 28 days by 5% to 10%, which coincides with the literature data [82–84], in which water-soluble sugars extracted from wood, in samples made without the addition of calcium chloride, are distributed throughout the volume of cement stone.

TABLE 2.13
Experimental Planning Conditions

	Factors			Variation Levels			
No.	Natural	Coded	−1	0	+1	Variation Interval
1	Sawdust/Cement ratio (S/C)	x_1	1:1.5	1:2	1:2.5	0.5
2	$CaCl_2$ content (%)	x_2	0	1	2	1

TABLE 2.14
Experimental-Statistical Models of the Water-Cement Ratio, Average Density, and Sawdust Concrete Strength

parameters	Statistical Models	
W/C	$W/C = 0.346 - 0.092x_1 + 0.007x_2$ $+ 0.035x_1^2 + 0.01x_2^2 - 0.005x_1x_2$	(2.38)
Average density (kg/m³)	$\rho = 1742 + 135x_1 + 35x_2 + 99x_1^2$ $- 2.5x_2^2 - 5.3x_1x_2$	(2.39)
Compressive strength at 7 days (MPa)	$f_c^7 = 7.4 + 2.9x_1 + 1.6x_2$ $+ 1.1x_1^2 + 0.1x_2^2 + 0.8x_1x_2$	(2.40)
Compressive strength at 28 days (MPa)	$f_c^{28} = 14.8 + 3.1x_1 0.4x_2 - 0.2x_1^2 + 0.2x_2^2$	(2.41)

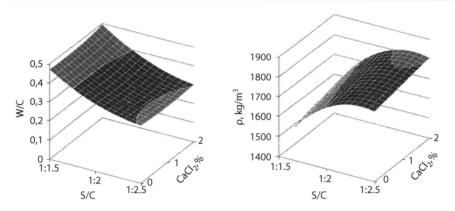

FIGURE 2.18 Influence of technological factors on the sawdust concrete, W/C, and average density.

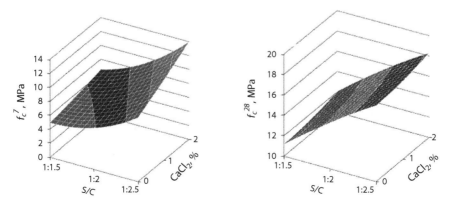

FIGURE 2.19 Influence of technological factors on the sawdust concrete strength.

In the samples with the addition of calcium chloride, they are blocked and do not hinder of the formation of cement stone.

Thus, the studies performed showed the possibility of obtaining structural and thermal insulation concretes for the manufacture of wall building materials, using LSC with clinker content less than 20% and wood processing industry waste, an organic aggregate, as a binder.

2.3 FIBER-REINFORCED CONCRETE BASED ON ACTIVATED LSC

2.3.1 Fiber Concrete at LSC

Construction of responsible structures for use in extreme conditions requires the use of effective concrete [20,52,63,74,85], which ensures high reliability and durability. Dispersion-reinforced high-strength fiber-reinforced concrete belongs to such effective concretes [20,85,86]. Dispersed fiber reinforcement makes it possible to compensate for the main disadvantages of cement concrete, namely, low tensile and bending strength, to reduce shrinkage and creep [20,58,87,88].

The objective of the work at this stage of the research was to study the complex effect on the strength of fiber-reinforced concrete made based on activated LSC, the consumption of cement, fiber, and superplasticizer additives with different workability of the concrete mixture.

Basalt fiber at 12 mm long was used as raw material for the experimental studies. The amount of fiber introduced was 0.5 and 1 kg/m^3 of concrete mix. Activated LSC with a clinker content of 12% was used as the binder. The strength of the binder with a specific surface area of 450 m^2/kg was 43 MPa. The plasticizing additive used was the superplasticizer SP-1. The aggregates for concrete were the granite-crushed stone with a maximum grain size of 20 mm and sand with $M_f = 1.9$.

The main studies were performed using a mathematical planning of the experiment. For this, a three-level four-factor plan was implemented [40]. The experimental planning conditions are given in Table 2.15.

After processing and statistical analysis of the experimental data, mathematical models of the water demand of the concrete mixture and the compressive and bending strength standard concrete samples based on the studied binders in the form of polynomial regression equations were obtained (Table 2.16).

TABLE 2.15
Experiment Planning Conditions

No.	Factors		Levels of Variation			
	Natural	Coded	−1	0	+1	Interval
1	Binder consumption (kg/m^3) (C)	x_1	300	400	500	100
2	Superplasticizer consumption (kg/m^3) (SP-1)	x_2	0	1.5	3	1.5
3	Fiber consumption (kg/m^3) (F)	x_3	0	0.5	1	0.5
4	Slump (cm) (S)	x_4	2	6	10	4

TABLE 2.16
Experimental-Statistical Models of Water Demand and Strength of Fiber Concrete at Activated LSC

Water requirement of concrete mix (L/m³)	$W = 187.4 + 1.2x_1 - 13x_2 + 0.84x_3 + 29.1x_4$ $+ 8.8x_1^2 - 2.2x_2^2 - 9.2x_3^2 - 4.2x_4^2 - 0.4x_2x_4$	(2.42)
Compressive strength at 7 days (MPa)	$f_c^7 = 20 + 3.1x_1 + 5.1x_2 + 0.3x_3 - 5.4x_4$ $- 6.4x_1^2 - 6.3x_2^2 - 10.8x_3^2 + 19.7x_4^2 + 1.9x_1x_2$ $- 0.8x_1x_3 - 2.4x_1x_4 - x_2x_3 - 2.2x_2x_4 + 0.8x_2x_4$	(2.43)
Compressive strength at 28 days (MPa)	$f_c^{28} = 24.7 + 9.9x_1 + 4x_2 + 0.54x_3 - 2.6x_4 - 4.5x_1^2$ $- 1.6x_2^2 - 8.8x_3^2 + 16.8x_4^2 + 2.6x_1x_2 + 0.06x_1x_3$ $- 2.6x_1x_4 + 0.3x_2x_3 - 0.5x_2x_4 - 0.15x_3x_4$	(2.44)
Bending strength at 28 days (MPa)	$f_{c,tb}^{28} = 3.01 + 1.21x_1 + 0.5x_2 + 0.63x_3 - 0.66x_4$ $- 0.098x_1^2 - 0.011x_2^2 - 0.0262x_3^2 - 0.13x_4^2$ $+ 0.34x_1x_2 + 0.41x_1x_3 - 0.292x_1x_4$ $+ 0.32x_2x_3 - 0.068x_2x_4 - 0.134x_3x_4$	(2.45)

Note: In equations (2.42) through (2.45):

$$x_1 = \frac{(C-400)}{100}; x_2 = \frac{(SP-1.5)}{1.5}; x_3 = \frac{(F-0.5)}{0.5}; x_4 = \frac{(S-6)}{6}.$$

The analysis of the obtained experimental-statistical models made it possible to determine the optimal levels of variable factors at which the best indicators of concrete strength are provided. For basalt fiber, they are: $x_1 = 0.84$ ($C = 484$ kg/m³), $x_2 = 0.91$ (SP-1 = 2.87 kg/m³), $x_3 = 0.53$ ($F = 0.77$ kg/m³), and $x_4 = 0.83$ ($S = 1$ to 2 cm), while providing compressive strength within 40 MPa and bending strength within about 7 MPa. Graphic dependencies illustrating the influence of technological factors on concrete compressive and bending strength are presented in Figures 2.20 and 2.21.

Analysis of the obtained experimental-statistical models indicates workability is the most significant factor affecting the water demand of the concrete mixture. A change in slump cone from 1 to 4 cm to 10 to 15 cm leads to an increase in water demand by an average of 30%. The influence of this factor is almost linear in nature and it accounts for about 70% of the influence of all other factors considered (Table 2.15). This effect can be somewhat mitigated by increasing the consumption of the SP-1 superplasticizer, the growth of which, of a value of 3 kg/m³, reduces the amount of water required to obtain a concrete mix with a given mobility by 15% and, accordingly, to increase the strength. An increase in the amount of binder, in varying limits, does not significantly affect the water demand of fiber-reinforced concrete. A certain increase in the amount

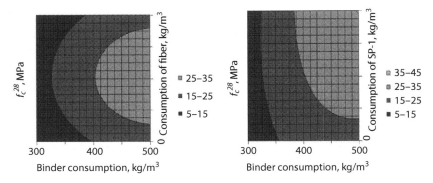

FIGURE 2.20 The influence of technological factors on the compressive strength of concrete at 28 days.

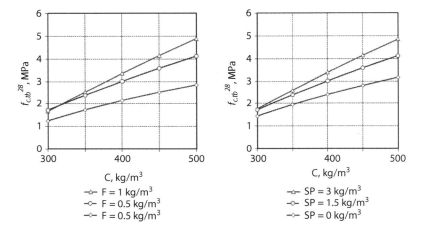

FIGURE 2.21 The influence of technological factors on the bending strength of fiber concrete.

of water required to obtain a concrete mixture with a given workability index causes an increase in fiber consumption to 0.5 kg/m³ of concrete mixture.

With an increase in binder consumption from 300 to 500 kg/m³ and consumption of superplasticizer from 0 to 3 kg/m³, the strength of fiber concrete increases by about 40%. An increase in the fiber content, but up to a value of not more than 0.5 kg/m³ of concrete mixture, also positively affects strength.

Analysis of the graphical dependencies shown in Figure 2.21 indicates the increase in fiber content is the most significant factor that affects the tensile strength of concrete in bending. Due to the three-dimensional dispersed reinforcement of the cement matrix of concrete with basalt fiber, the bending strength of concrete is almost doubled.

The second stage of research was devoted to establishing the influence of the amount of basalt fiber on the strength characteristics of fine-grained concrete. The influence of basalt fiber at 12 and 24 mm long also was compared. In the course of research, at each point of the plan, fine-grained concrete with composition of a

Activated Low Clinker Slag Portland Cement and Concrete on Its Basis

binder:sand at 1:3 was made. The W/C was determined to achieve the spread of the solution cone on the shaking table in the range of 112 to 115 mm, as well as the compressive and bending strength of the beam samples at 7 and 28 days.

Studies have established that the use of basalt fiber in the composition of fine-grained concrete made with activated LSC allows an increase its compressive strength by 14% and bending strength by 35%. An increase in fiber length also does not significantly affect the properties of fiber concrete, and a somewhat worse performance of concrete with basalt fiber of 24 mm long compared to 12 mm fiber (Table 2.17) can be explained by the fact that longer fibers are less distributed in the thickness of concrete. An increase in the amount of basalt fiber over 1 kg/m³ negatively affects the strength characteristics of fiber concrete. This result is due to an increase in the amount of water necessary to obtain concrete mixtures with a given mobility.

Based on the obtained experimental-statistical model (2.45) (Table 2.16), a nomogram of the strength of fiber-reinforced concrete fabricated on activated LSC was constructed (Figure 2.22). This nomogram predicts the strength of concrete with a given composition or determines the consumption of one of the components for a given content of the others. It, together with the complex of the obtained models (Table 2.16), can be used also for designing fiber concrete compositions with a set of specified properties. The calculation procedure is as follows:

1. Using the nomogram shown in Figure 2.22, determine the consumptions of the binder, superplasticizer SP-1, and fiber, providing the necessary bending strength of concrete with a given workability.
2. Translate the obtained values of the components consumptions into the coded form using formulas 2.46 through 2.49:

$$x_1 = \frac{(C-400)}{100}; \quad (2.46)$$

$$x_2 = \frac{(SP-1-1.5)}{1.5}; \quad (2.47)$$

TABLE 2.17
The Effect of Basalt Fibers of Various Lengths on the Properties of Fine-Grained Concrete

No.	Fiber Consumption (kg/m³)	24-mm Fiber				12-mm Fiber			
		W/C	$f_{c.tb}^{7}$ (MPa)	f_{c}^{28} (MPa)	$f_{c.tb}^{28}$ (MPa)	W/C	$f_{c.tb}^{7}$ (MPa)	f_{c}^{28} (MPa)	$f_{c.tb}^{28}$ (MPa)
1	0	0.34	1.82	22.6	4.9	0.34	1.82	22.6	4.9
2	1.0	0.39	1.9	23.4	7.1	0.35	6.4	26.2	7.6
3	5.0	0.39	2.65	19.6	6.8	0.39	5.3	22.6	5.8
4	10.0	0.42	2.56	18.5	6.5	0.39	5.6	14.7	5.6
5	15.0	0.44	3.35	17.6	5.8	0.4	4	16.5	5.3

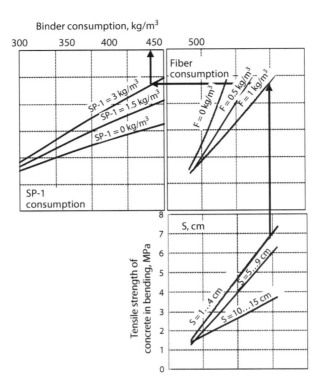

FIGURE 2.22 Nomogram of bending strength of fiber reinforced concrete manufactured at activated LSC.

$$x_3 = \frac{(F-0.5)}{0.5}; \qquad (2.48)$$

$$x_4 = \frac{(S-6)}{6}. \qquad (2.49)$$

3. Substituting the obtained values in equation (2.44), checking this component composition will provide the necessary compressive strength of concrete at 28 days. If durability is not provided, measures must be taken to reduce the W/C of the concrete mix.
4. Substitute the transformed value into the coded form of the value of the consumption of materials in equation (2.42), and determine the consumption of water that will provide the necessary mobility of the concrete mixture and the specified strength characteristics of concrete.
5. Having determined the consumption of cement and water consumption, according to the well-known formulas 2.32 and 2.33, the consumptions of aggregates are determined.

2.3.2 CALCULATION EXAMPLE

Calculate the composition of fiber-reinforced concrete manufactured at activated LSC with a 28-day compressive strength of 25 MPa and bending strength of 6 MPa. The slump cone of the concrete mix is 6 cm. The superplasticizer SP-1 is used as a plasticizing additive. The coefficient $\alpha = 1.46$, bulk density of crushed stone $\rho_{b.cr.st} = 1.65$ g/cm^3, density of crushed stone $\rho_{cr.st} = 2.85$ g/cm^3, and density of sand $\rho_s = 2.65$ g/cm^3 [31]

1. According to the nomogram shown in Figure 2.22, determine the consumption of the binder, superplasticizer SP-1, and fiber, ensuring compliance with the requirements ($f_c^{28} = 30$ MPa, $f_c^{28} = 6$ MPa, $S = 6$ cm).
2. Transform the obtained values ($C = 485$ kg/m^3, SP-1 $= 3$ kg/m^3, $F = 1$ kg/m^3, $S = 6$ cm) in coded form:

$$x_1 = \frac{(C-400)}{100} = \frac{(485-400)}{100} = 0.85;$$

$$x_2 = \frac{(SP-3-0.3)}{0.3} = \frac{(3-1.5)}{1.5} = 1;$$

$$x_3 = \frac{(F-0.5)}{0.5} = \frac{(1-0.5)}{0.5} = 1;$$

$$x_4 = \frac{(S-6)}{6} = \frac{(6-6)}{6} = 0.$$

3. Substituting the results in equation (2.44), checking this composition will provide the necessary compressive strength of concrete at 28 days ($f_c^{28} \geq 25$ MPa).

$f_c = 24.7 + 9.9 \times 0.85 + 4 \times 1 + 0.54 \times 1 - 2.6 \times 0 - 4.5 \times 0.85^2$

$- 1.6 \times 1^2 - 8.8 \times 1^2 + 16.8 \times 0^2 + 2.6 \times 0.85 \times 1 + 0.06 \times 0.85 \times 1$

$- 2.6 \times 0.85 \times 0 + 0.3 \times 1 \times 1 - 0.5 \times 1 \times 0 - 0.15 \times 1 \times 0 = 26.6$ MPa,

$- 26.6 \geq 25$ the condition performed.

4. Substitute, determined by the nomogram of material consumptions, into equation (2.42). Previously converting them to coded values using formulas (2.46) through (2.49), determine the water consumption that will provide the necessary mobility of the concrete mixture and the specified strength characteristics of concrete:

$W = 187.4 + 1.2 \times 0.85 - 13 \times 1 + 0.84 \times 1 + 29.1 \times 0 + 8.8 \times 0.85^2$

$- 2.2 \times 1^2 - 9.2 \times 1^2 - 4.2 \times 0^2 - 0.4 \times 1 \times 0 = 171$ L/m^3.

5. Having determined the consumption of cement and water consumption, according to the formulas (2.32) and (2.33), determine the consumptions of aggregates:

$$CS = \frac{1000}{\alpha \frac{P_{c.s}}{\rho_{b.c.s}} + \frac{1}{\rho_{c.s}}} = \frac{1000}{1.46 \frac{0.42}{1.65} + \frac{1}{2.85}} = 1388 \text{ kg/m}^3.$$

$$Sd = \left(1000 - \left(\frac{C}{\rho_c} + W + \frac{CS}{\rho_{c.s}}\right)\right) \times \rho_s$$

$$= \left(1000 - \left(\frac{485}{3.1} + 171 + \frac{1388}{2.85}\right)\right) \times 2.65 = 492 \text{ kg/m}^3.$$

Estimated composition of concrete: cement, 485 kg/m³; water, 171 L/m³; crushed stone, 1388 kg/m³; and sand, 492 kg/m³. The consumption of SP-1 superplasticizer is 3 kg/m³ and the consumption of basalt fiber is 1 kg/m³.

For a comparative analysis, studies were conducted to establish the effect of polypropylene fiber on the strength characteristics of fiber concrete manufactured at activated LSC. For this, planned experiments were carried out for three factors according to plan B_3. The experimental planning conditions are given in Table 2.18.

The results of processing and statistical analysis of experimental data are given in Table 2.19.

Graphic dependences illustrating the influence of technological factors on the compressive and bending strength of concrete are shown in Figures 2.23 and 2.24.

Analysis of the regression equations (Table 2.19) and graphical dependencies shown in Figure 2.23 indicates the strength of fiber concrete made using polypropylene fibers as well as the strength of basalt fiber concrete are increased with increasing consumption of binder and the addition of superplasticizer SP-1. It is associated with a decrease in W/C. It is also worth noting that the influence of these factors is the largest and their contribution to 28 days of compressive strength of concrete is 60% and 20%, respectively. The increase in the content of polypropylene fiber, which, in comparison with basalt fiber, is better distributed by volume of concrete, does not significantly affect the strength of concrete made based on LSC, which is associated with the worse adhesive properties of polypropylene fibers.

TABLE 2.18
Experiment Planning Conditions

| No. | Factors | | Variation Levels | | | |
	Natural	Coded	−1	0	+1	Interval
1	Binder consumption (kg/m³) (C)	x_1	300	400	500	100
2	Plasticizer consumption (kg/m³) (SP-1)	x_2	0	1.5	3	1.5
4	Fiber consumption (kg/m³) (F)	x_3	0	0.5	1	0.5

TABLE 2.19
Experimental-Statistical Models of Fiber Concrete Strength

Source Parameter	Statistical Models	
Compressive strength at 28 days (MPa)	$f_c^{28} = 15.6 + 9x_1 + 3.4x_2 + 0.81x_3$ $\quad + 7.4x_1^2 - 2x_2^2 + 1.3x_3^2 + 2.8x_1x_2$ $\quad - 0.8x_1x_3 + 0.3x_2x_3$	(2.50)
Bending strength at 28 days (MPa)	$f_{c,bt}^{28} = 1.95 + 0.7x_1 + 0.2x_2 + 0.15x_3$ $\quad + 0.32x_1^2 - 0.26x_2^2 + 0.42x_3^2 + 0,13x_1x_2$ $\quad + 0.05x_1x_3 + 0.19x_2x_3$	(2.51)

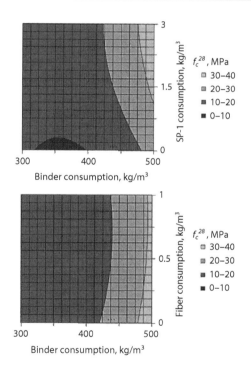

FIGURE 2.23 Influence of technological factors on the compressive strength of concrete reinforced with polypropylene fiber at 28 days.

Analysis of the graphic dependencies in Figure 2.24 it indicate when polypropylene fibers are used as reinforcing fibers, the effect is slightly worse in comparison with basalt fibers. This result may be due to poor adhesion of the polypropylene with the cement matrix. However, an increase in the content of these fibers makes it possible to increase the bending strength of concrete by 10% to 15%, which coincides with the data [89,90].

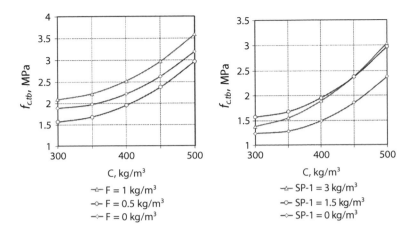

FIGURE 2.24 The influence of technological factors on the bending strength of concrete reinforced with polypropylene fiber at 28 days.

Thus, studies have shown the possibility of almost doubling the bending strength of concrete using basalt fiber with a fiber length of 12 mm. This effect is achieved due to the three-dimensional reinforcement of the cement matrix. It was also found that an increase in the length of basalt fiber from 12 to 24 mm does not significantly affect the properties of fiber-reinforced concrete manufactured on low-clinker cement.

2.4 FROST RESISTANCE OF CONCRETE AND FIBER-REINFORCED CONCRETE AND CORROSION RESISTANCE IN CONCRETE AT LSC

2.4.1 Frost Resistance of Concrete and Fiber-Reinforced Concrete Made Based on LSC

For normal-weight concrete at LSC, operated in the cold season, when multiple transitions through 0°C are characteristic, frost resistance is one of the most important properties that determine its durability. According to modern concepts, the main factors determining the frost resistance of concrete are the parameters of its pore space, which determine the degree of saturation of this concrete with water, the kinetics of ice formation in pores, and the intensity of destructive stresses arising and developing in concrete [91–93]. The modern theory of concrete frost resistance is based on the fundamental works according to which frost resistance is ensured, first, by creating a system of conditionally closed pores that compensate for internal deformations during freezing and thawing.

Concrete on slag Portland cement are characterized by less resistance to frost than on conventional Portland cement. Therefore, the task of the given stage of research was to establish possible ways to increase the frost resistance of concrete on activated LSC. To accomplish this task, experiments were carried out according to plan B_4 [40]. The conditions for the planning of experiments and the results of experimental studies are given in Tables 2.20 and 2.21.

TABLE 2.20
Experiment Planning Conditions

No.	Factors		Variation Levels			
	Natural	Coded	−1	0	+1	Interval Varying
1	Binder consumption (kg/m³)	x_1	300	400	500	100
2	Plasticizer consumption (kg/m³)	x_2	0	1.5	3	1.5
3	AEA consumption (kg/m³)	x_3	0	0.025	0.05	0.005
4	Fiber consumption (kg/m³)	x_4	0	0.5	1	0.5

TABLE 2.21
The Results of Experimental Studies

No.	C (kg/m³)	W (L)	Sd, (kg/m³)	Cr. St, (kg/m³)	SP-1, (kg/m³)	Fiber (kg/m³)	AEA (%)	S (cm)	f_c^7 (MPa)	f_c^{28} (MPa)	F (Cycle)
1	500	171	740	1014	3	1	0.05	5	12.8	27.6	210
2	500	171	740	1014	3	0	0.05	9	17.7	28.1	210
3	500	171	740	1014	3	1	0	5	26	38.2	210
4	500	171	740	1014	3	0	0	5	26.3	40.7	210
5	500	198	711	974	0	1	0.05	5.5	19.5	28.7	135
6	500	198	711	974	0	0	0.05	5	14.9	22.2	135
7	500	198	711	974	0	1	0	5	11.6	25.8	135
8	500	198	711	974	0	0	0	6	11.9	25.4	120
9	300	171	811	1112	3	1	0.05	9	6.2	15.1	165
10	300	171	811	1112	3	0	0.05	9	4.5	9.4	30
11	300	171	811	1112	3	1	0	6	7.3	14.6	150
12	300	171	811	1112	3	0	0	9	6.4	12.9	105
13	300	198	782	1072	0	1	0.05	6	5.3	9.8	135
14	300	198	782	1072	0	0	0.05	9	4.5	8	90
15	300	198	782	1072	0	1	0	7.5	5.7	8.5	135
16	300	198	782	1072	0	0	0	5.5	5.3	14.1	120
17	500	194	715	980	1.5	0.5	0.025	5	15.4	30.8	240
18	300	186	796	1090	1.5	0.5	0.025	7	7.6	16.5	90
19	400	171	776	1063	3	0.5	0.025	7	7.2	20.2	90
20	400	198	746	1023	0	0.5	0.025	5	8.1	19.8	105
21	400	179	768	1052	1.5	0.5	0.05	6	12.7	21.5	135
22	400	186	760	1041	1.5	0.5	0	5	14.3	29.5	135
23	400	186	760	1041	1.5	1	0.025	5	11.2	21.6	165
24	400	200	744	1020	1.5	0	0.025	5	14.5	21.7	135

For conducting experimental studies, basalt fibers at 12 mm long were used. The amount of fiber introduced ranged from 0 to 1 kg/m³ of the concrete mix. An activated LSC of the following composition was used as a binder: clinker, 12%; blast furnace granulated slag, 88%; and phosphogypsum dihydrate, 7.5% (in terms of SO_3, 4.5%). In addition, lime (3%) and sodium silicon fluoride (2%) were added to the

TABLE 2.22
Experimental-Statistical Models of Concrete Strength and Frost Resistance on LSC

Parameters	Statistical Models	
Compressive strength at 28 days (MPa)	$f_c^{28} = 23.4 + 8.9x_1 + 2.5x_2 - 2.2x_3 + 0.4x_4$	
	$+ 0.2x_1^2 - 3.5x_2^2 + 2.0x_3^2 - 1.8x_4 + 1.3x_1x_2 - x_1x_3$	(2.52)
	$+ 0.02x_1x_4 - 1.3x_2x_3 + 0.1x_2x_4 + 1.2x_3x_4$	
Frost resistance, cycle	$f_{r.c} = 134.8 + 32.8x_1 + 15.1x_2 + 4.2x_3 + 16x_4$	
	$+ 30x_1^2 - 37.5x_2^2 - 0.03x_3^2 + 15x_4^2 + 21.6x_1x_2$	(2.53)
	$+ 6.6x_1x_3 - 14.1x_1x_4 - 2.8x_2x_3 + 6.6x_2x_4$	
	$+ 6.6x_3x_4$	

binder. The superplasticizer SP-1 was used as a plasticizing additive. The aggregates for concrete were granite-crushed stone with a maximum grain size of 20 mm and sand with $M_f = 1.9$. Determination of frost resistance was carried out by the accelerated method.

An analysis of the obtained equations (Table 2.22) and graphical dependencies (Figure 2.25) indicates that all four factors significantly affect the frost resistance of concrete. With an increase in their values, the frost resistance of concrete at the LSC grows, and according to the effect on the strength, they can be arranged in the row: $x_1 > x_4 > x_2 > x_3$.

An increase in the content of the binder and the additive of the superplasticizer SP-1 leads to a sharp increase in the frost resistance of concrete, which is associated with a decrease in the W/C and, accordingly, a decrease in the open porosity of concrete. Also, frost resistance is positively influenced by an increase in the content of air-entraining additive and fiber consumption. Due to the three-dimensional reinforcement of the cement matrix with basalt fiber, the tensile strength of concrete increases, which leads to an increase in the resistance of concrete to cyclic stresses that occur in concrete when it is alternately frozen and thawed.

Using the superplasticizer SP-1, basalt fiber, and the air-entraining additive (AEA), concrete grades with frost resistance F200-F250 were obtained, while the compressive strength of concrete ranged from 30 to 40 MPa. Such frost resistance of concrete allows for its operation in harsh conditions.

2.4.2 Corrosion Resistance in Concrete at LSC

Corrosion resistance of reinforcement in reinforced concrete structures is based on the high quality of concrete, its density, and strength. These qualities provide enough conditions for the preservation of reinforcing steel in an environment with a pH of 12 to 13. Steel is coated with a reliable passive layer at 2 to 20 nm thick [69]. The low

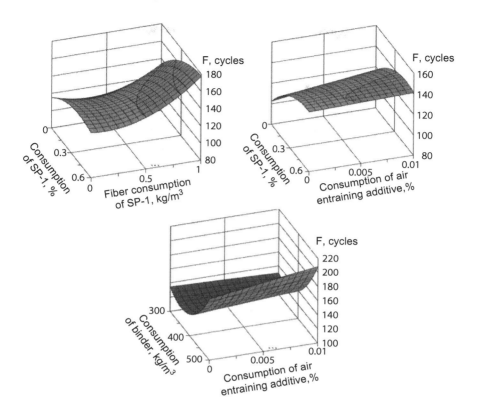

FIGURE 2.25 The influence of variable factors on the frost resistance of concrete made at activated LSC.

clinker content in non-activated LSC causes low pH values (the pH of the studied binders is about 10) of concrete and mortar mixtures; therefore, concrete on this cement does not provide passivation of reinforcing steel, which leads to corrosion and the gradual destruction of reinforcement. The use of LSC in structural products requires the use of anti-corrosion coatings for reinforcement, which significantly increases the cost of construction.

A series of experiments was performed in which samples of reinforcing bars with smooth wire at a diameter of Ø 6 mm of class B-I were located inside samples of beams from a mortar based on LSC (beams of the composition LSC:sand = 1:3) of different composition: without additives and with additives of SP-1 and corrosion inhibitor reinforcement—sodium nitrite ($NaNO_2$) (Table 2.23). For comparison, samples of steel reinforcement were placed inside the samples of beams made with the activated LSC. Moreover, during mixing of the mortar mixture, no additives of superplasticizer and corrosion inhibitor were used.

Concrete samples with reinforcement were stored in water, in air, and in a solution of sodium chloride at an ambient temperature of 20°C ± 2°C.

TABLE 2.23
Mortar Compositions for Determination of Reinforced Steel Corrosion Resistance

No.	SP-1 (%)	NaNO$_2$	Cement (g)	Sand (g)	W/C
1	0.6	3	500	1500	0.32
2	0.6	0	500	1500	0.32
3	0	3	500	1500	0.39
4	0	0	500	1500	0.39
5	0.6	1.5	500	1500	0.32
6	0	1.5	500	1500	0.34
7	0.3	3	500	1500	0.34
8	0.3	0	500	1500	0.34
9	0.3	1.5	500	1500	0.34

As a criterion for corrosion of the reinforcement, the presence and intensity of rust on the surface of the reinforcement was evaluated. The storage duration of the samples was 5 months (150 days) (Figures 2.26 through 2.36).

According to the results obtained, when samples were stored in air, there were no signs of corrosion of the reinforcement in concrete on inactive LSC. This result can probably be explained by the low humidity of the air, as well as the high density of concrete, which made it impossible for moisture to penetrate through the thickness of the concrete to the reinforcement, and accordingly, conditions for intensifying the process of its corrosion were not provided.

FIGURE 2.26 Corrosion of reinforcing bars in samples manufactured at non-activated LSC without additives stored in air.

FIGURE 2.27 Corrosion of reinforcing bars in samples manufactured at non-activated LSC, with additives SP-1 (0.3%) and NaNO$_2$ (1.5%) stored in air.

Activated Low Clinker Slag Portland Cement and Concrete on Its Basis   111

FIGURE 2.28 Corrosion of reinforcing bars in samples manufactured at non-activated LSC without additives stored in water.

FIGURE 2.29 Corrosion of reinforcement in samples manufactured at non-activated LSC, with additives SP-1 (0.3%) and $NaNO_2$ (1.5%) stored in water.

FIGURE 2.30 Corrosion of reinforcing bars in samples manufactured at non-activated LSC without additives stored in a solution of sodium chloride.

FIGURE 2.31 Corrosion of reinforcing bars in samples manufactured at non-activated LSC, with additives SP-1 (0.6%) and $NaNO_2$ (1.5%) stored in a solution of sodium chloride.

FIGURE 2.32 Corrosion of reinforcing bars in samples manufactured at non-activated LSC, with the addition of SP-1 (0.6%) stored in a solution of sodium chloride.

FIGURE 2.33 Corrosion of reinforcing bars in samples manufactured at non-activated LSC, with the addition of SP-1 (0.3%) stored in a solution of sodium chloride.

FIGURE 2.34 Corrosion of reinforcing bars in samples manufactured at activated LSC stored in air.

FIGURE 2.35 Corrosion of reinforcing bars in samples manufactured at activated LSC stored in water.

Activated Low Clinker Slag Portland Cement and Concrete on Its Basis

FIGURE 2.36 Corrosion of reinforcing bars in samples manufactured at activated LSC stored in a solution of sodium chloride.

Visible manifestations of corrosion are observed in samples that were stored in a solution of sodium chloride and in water. Significant reinforcement corrosion rate is typical for concrete samples made without additives. This result can be explained by the fact that an increase in the water demand of concrete made without the use of plasticizing additives leads to an increase in its porosity and, accordingly, a decrease in water resistance.

A decrease in W/C to 0.32 to 0.34 due to the use of superplasticizer SP-1 in the amount of 0.6% and 0.3% by weight of the binder, and a corresponding significant increase in density and a decrease in the porosity of concrete significantly reduced corrosion damage fittings—only the tip of the wire, not covered by concrete, substantially rusted. In this case, the corrosion rate of the metal covered with concrete was insignificant—there were separate small spots of rust, the amount of which was approximately the same for samples with different SP-1 content in the binder.

The corrosion rate of reinforcing bars in concrete at non-activated LSC during storage in water turned out to be almost the same as when stored in a solution of sodium chloride. Perhaps this result can be explained by the test conditions: in both cases, the reinforcement samples after formation were exposed to the aqueous medium of the cement stone. When stored in water in a cement stone, a sufficiently high level of humidity was maintained, which contributed to the corrosion of the metal. When testing the samples in a solution of sodium chloride, a slightly concentrated liquid was probably made and a sufficiently aggressive environment was not provided, therefore, corrosion showed itself in almost the same way as in water. The use of SP-1 additive also made it possible to significantly reduce the corrosion damage of reinforcement by increasing the density and decreasing porosity of concrete. A significant effect on the corrosion process, an increase in the content of SP-1 in these experiments, was not noted.

Separate use of a steel corrosion inhibitor additive ($NaNO_2$) without compaction of the concrete structure due to the SP-1 superplasticizer also gave a certain effect of reducing corrosion, compared with the reinforcement samples, which were stored in concrete without additives. The simultaneous introduction of superplasticizer SP-1 in mortar mix at W/C = 0.32 and steel corrosion inhibitor $NaNO_2$ in an amount of 3% by weight of the binder made it possible to reduce reinforcement damage to barely noticeable traces of corrosion, but impossible to stop reinforcement corrosion completely in concrete in inactive LSC using additives.

According to literature sources [7,19,39], steel reinforcement in concretes made based on low-clinker cements is capable of resisting corrosion for a long time when such concrete is used only in an air-dry environment ($W < 60\%$); in an aggressive environment, and also in water, steel reinforcement in this concrete corrodes. The main cause of corrosion of the reinforcement is the low alkalinity of the cement stone environment in non-activated LSC.

However, studies have established the possibility of increasing the alkalinity of the LSC medium by additionally introducing 3% binder lime into the composition and adding 2% Na_2SiF_6 hardening activator sodium silicon fluoride. At the same time, not only the activity of the binder due to sulfate-fluoride-alkaline activation is growing, but also the pH of the medium. Studies have established that the pH of activated LSC is 12 to 13, and this environment provides reliable protection for steel reinforcement, which was confirmed by our research. When using activated LSC as a binder, corrosion of reinforcing steel is not observed either when storing samples in air, or when storing them in water or a solution of sodium chloride.

2.5 DRY BUILDING MIXES AND MORTARS BASED ON ACTIVATED LSC

The use of cement with a high content of active mineral additives, including LSC, is promising in the production of dry building mixes in compositions with the necessary modifier additives [94–96].

2.5.1 Dry Mixes for Masonry Mortars

The complex effect on the properties of masonry mortars based on dry construction mixtures made using activated LSC, cement, lime, and plasticizing additives was investigated.

As raw materials for obtaining dry mixtures and mortars based on them, LSC was used with a clinker content of 12% and an activity of 43 MPa. As a plasticizing additive, an LST was used. Quartz sand with a maximum grain size of 1.25 mm and fineness modulus of $M_f = 1.9$ was used as a fine aggregate.

The experimental planning conditions are given in Table 2.24. All samples of the mortar mixture were made with the same mobility of 4 to 8 cm by immersion of a standard cone [96].

The experimental statistical models in coded variables of water demand, water-retention capacity, availability time of the mortar mixture, and the compressive strength of the mortar in the form of polynomial regression equations are given in Table 2.25.

TABLE 2.24
Experimental Planning Conditions

No.	Factors		Variation Levels			Interval
	Natural	Coded	−1	0	+1	
1	Cement consumption (C) (kg/m³)	x_1	200	300	400	100
2	Plasticizing additives (LST) consumption (kg/m³)	x_2	1	1.5	2	0.5
3	Lime consumption (L) (kg/m³)	x_3	10	25	40	15

TABLE 2.25
Experimental-Statistical Models of Concrete, Mix Properties on Activated LSC

Parameters	Statistical Models	
Water demand of solution mixture (L/m³)	$W = 145.2 + 6.4x_1 6.02x_2 + 7.9x_3$ $+ 6.6x_1^2 + 0.4x_2^2 + 0.3x_3^2 + 0.46x_1x_2$ $+ 4.6x_1x_3 + 0.7x_2x_3$	(2.54)
Water-retention capacity (%)	$W_{ret} = 98.7 + 0.19x_1 + 0.16x_2 + 0.57x_3$ $- 0.26x_1^2 - 0.1x_2^2 + 0.16x_3^2 + 0.03x_1x_2$ $- 0.32x_1x_3 - 0.27x_2x_3$	(2.55)
Availability of mortar mix (min)	$\tau = 179.5 - 46.2x_1 - 0.1x_2 + 14.8x_3$ $+ 7.2x_1^2 - 1.3x_2^2 - 1.8x_3^2 - 0.1x_1x_2$ $- 3.9x_1x_3 + 3.4x_2x_3$	(2.56)
Compressive strength of mortar at 7 days (MPa)	$f_c^7 = 6.33 + 1.9x_1 + 1.0x_2 - 1.49x_3$ $+ 1.9x_1^2 + 0.5x_2^2 + 0.8x_3^2 - 1.7x_1x_2$ $- 0.8x_1x_3 - 0.06x_2x_3$	(2.57)
Compressive strength of mortar at 28 days (MPa)	$f_c^{28} = 12.6 + 6x_1 + 2.7x_2 - 1.2x_3$ $+ 3.1x_1^2 + 0.3x_2^2 - 0.01x_3^2 + 1.1x_1x_2$ $+ 0.04x_1x_3 - 0.26x_2x_3$	(2.58)

The dependences obtained by the analysis of models illustrating the influence of technological factors on the water demand of the mortar mixture and the compressive strength of the mortar in 28 days are presented in Figures 2.37 through 2.39. From them follows that an increase in cement consumption within a variable range leads to some increase in the water demand of the mortar, but this effect is insignificant.

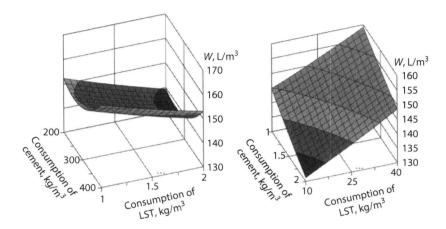

FIGURE 2.37 Influence of technological factors on water demand of masonry mortar.

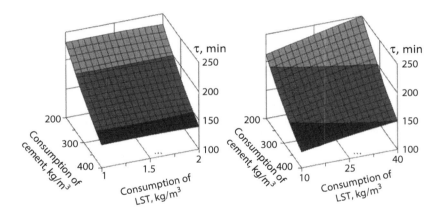

FIGURE 2.38 Influence of technological factors on the shelf life of masonry mortar.

This result leads to the assumption that in the mortar of the LSC, as well as in the concrete, there is a known rule of constancy of water demand [96]. Significant impact on water demand is an increase in consumption of hydrated lime. Thus, increasing the lime content to 40 kg/m^3 leads to an increase in water demand of the mixture more than 1.5 times. However, this water demand can be offset by adjusting the content of the plasticizing additive LST, which reduces the amount of water required to obtain a plastic mortar with a given mobility and, accordingly, to ensure its ultimate strength. The influence of these two factors on the water demand of the mortar is the most significant and is almost linear (Figure 2.37).

Important technological properties of mortar mixtures intended for masonry work are enough water-retention ability and, as a result, low water separation. The latter, as is known [56], is caused by sedimentation processes in a freshly laid

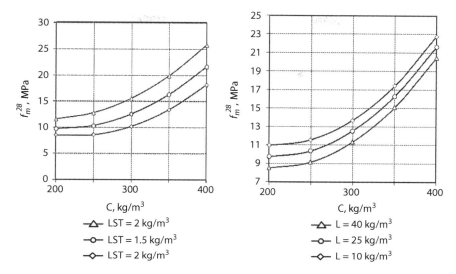

FIGURE 2.39 Influence of technological factors on the strength of masonry mortar at 28 days.

mortar mixture. Sedimentation processes can take place in the period preceding the formation of a stable coagulation structure. In practice, water separation is significantly reduced by introducing lime dough and dispersed hydraulic additives into mortar mixtures.

From equation (2.55) (Table 2.25) it follows that all three factors affect the water-retention ability of masonry mortars made based on activated LSC. With an increase in their values, the water-retention ability increases. By influence, factors can be arranged in a row: $x_3 > x_1 > x_2$. It is also worth noting that the water-retention capacity of mortar in the entire range of variation of factors is more than 95%. W/C masonry mortars made at activated LSC without using hydrated lime additive was in the range W/C ≈ 1.65 NC, where NC normal consistency cement. As shown by I. N. Akhverdov [43], at such W/C values, enough water-retention capacity of the cement paste is preserved.

In the technology of mortars, the safety of their mobility over time is becoming increasingly important. The introduction of mineral additives to cement is accompanied by a slowdown in early structure formation and, accordingly, extension of the time availability of mortar [96]. The analysis of the experimental-statistical model 2.56 (Figure 2.38) indicates the most significant effect on the time availability ("vitality") of mortars made based on LSC has a binder consumption. With an increase of binder consumption from 200 to 400 kg/m³, the time of use of the mortars is reduced possibly by almost half. This result is due to an increase in the amount of the active component of the mortars. A certain increase in the time availability of masonry mortars made at the LSC causes an increase in the content of hydrated lime, as well as the content of the plasticizing additive LST.

An increase in the strength of masonry mortars at 28 days leads to an increase in the consumption of binder and plasticizing additives, mainly due to a decrease in

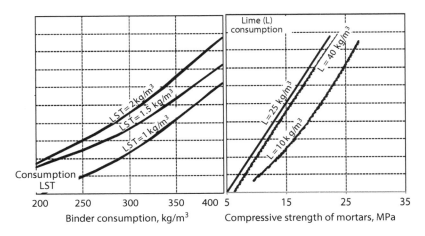

FIGURE 2.40 Strength nomogram of masonry mortars made based on LSC.

the W/C (Figure 2.39). In the range of compositions of mixtures, determined by the range of variation of the main factors (Table 2.24), it is possible to obtain mortars with a compressive strength of 5 MPa to 23 MPa. The strength of masonry mortars is negatively affected by an increase in the amount of hydrate lime, which is mainly associated with an increase in the amount of water necessary to obtain mortars with a given mobility.

Based on the obtained experimental-statistical model 2.58 (Table 2.25), a nomogram of the strength of masonry mortars was obtained (Figure 2.40). This nomogram predicts the strength of mortars with a given component composition or determines the consumption of one of the components for a given content of the others. It, together with a set of other models characterizing the properties of mortar mixtures (Table 2.25), also can be used to design mixes with a minimum cost and a set of specified properties.

2.5.2 Calculation Example

Calculate the composition of dry mixture for masonry mortars, which provides at a minimum the cost of obtaining mortar with a time availability of $\tau \geq 145$ minutes, compressive strength at 7 days $f_m^7 \geq 5$ MPa and 28 days $f_m^{28} = 15$ MPa.

Minimizing the cost of the mixture means minimizing the cost of the binder, to which not only lime, but also the plasticizer is conditionally applied:

$$C_{\text{bind}} = C_{cem} \times C + C_{LST} \times LCT + C_L \times L, \tag{2.59}$$

where C_{cem}, C_{LST}, and C_L, respectively the cost in euros per kilogram of cement, plasticizer, and lime and C, LST, L, respectively the consumption in kilograms per cubic meter of cement, plasticizer, and lime.

Activated Low Clinker Slag Portland Cement and Concrete on Its Basis

Provided that the required time availability and compressive strength of the mortar are provided at 7 and 28 days:

$$\tau \geq f_1(x_1, x_2, x_3), \quad (2.60)$$

$$f_m^{28} = f_2(x_1, x_2, x_3), \quad (2.61)$$

$$f_m^7 \geq f_3(x_1, x_2, x_3). \quad (2.62)$$

The methodology for calculating the cost-effective compositions of dry building mixtures using polynomial models and the linear programming method can be as follows:

1. The constraint functions (2.60) and (2.61) at $\tau = 145$ min and $f_m^{28} = 15$ MPa have the following form:

$$145 = 179.5 - 46.2x_1 - 0.1x_2 + 14.8x_3 + 7.2x_1^2 - 1.3x_2^2$$
$$- 1.8x_3^2 - 0.1x_1x_2 - 3.9x_1x_3 + 3.4x_2x_3;$$

$$15 = 12.6 + 6x_1 + 2.7x_2 - 1.2x_3 + 3.1x_1^2 + 0.3x_2^2 - 0.01x_3^2$$
$$+ 1.1x_1x_2 + 0.04x_1x_3 - 0.26x_2x_3;$$

2. Stabilize the factor x_2:
   a. At the basic level (LST = 1.5 kg/m³).
   b. At the maximum level (LST = 2 kg/m³).
   Then rewrite the previous equations in the form:
   a. $34.5 - 46.2x_1 + 14.8x_3 + 7.2x_1^2 - 1.8x_3^2 - 3.9x_1x_3 = 0;$

   $-2.4 + 6x_1 - 1.2x_3 + 3.1x_1^2 - 0.01x_3^2 + 0.04x_1x_3 = 0;$

   b. $33.1 - 46.2x_1 + 18.2x_3 + 7.2x_1^2 - 1.8x_3^2 - 3.9x_1x_3 = 0;$

   $0.6 + 7.1x_1 - 1.44x_3 + 3.1x_1^2 - 0.01x_3^2 + 0.04x_1x_3 = 0.$

3. Construct the isolines τ and f_m^{28} depending on the factors x_1 and x_3 and determine the range of possible solutions (the hatched section of Figure 2.41).
4. By the formula (2.61), find the cost of the binder at the points on the boundary of the experiment. Taking $Cc = 0.065$ Euro/kg, $C_{LST} = 14$ Euro/kg, $C_L = 0.18$ Euro/kg, calculate the value of the binder at the points at the intersection of the boundary of the region of possible solutions and the range of values of factors x_1 and x_3 (Figure 2.41). As can be seen from the figure, the minimum value of the binder will be at a point on the border of the allowable area with coordinates (−0.34; 1) belonging to the level line of

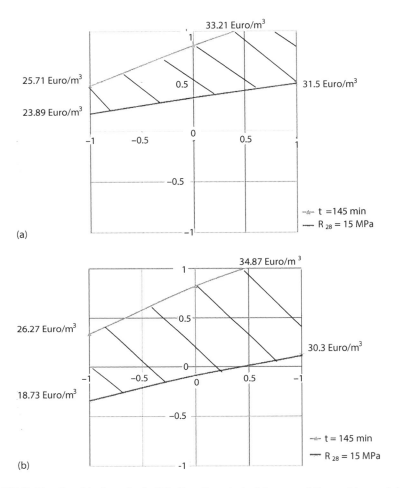

FIGURE 2.41 Graphical method of finding the admissible area of the problem solving at (a) $x_2 = 0$ and (b) $x_2 = 1$.

the function f_m^{28} (Figure 2.41b). Also from Figure 2.41, the decrease in the consumption of plasticizing additives LST from 2 to 1.5 kg/m³ entails an increase in binder consumption from 266 to 318 kg/m³ and a corresponding increase in the cost of dry mix by 5.16 Euro/m³.

5. Substitute the obtained values of the factors $x_1 = -0.34$, $x_2 = 1$, $x_3 = 1$ in equation (2.59) of Table 2.25 and check the fulfillment of the condition $f_m^7 \geq 5$ MPa, which gives:

$f_m^7 = 9.9$ MPa ≥ 5 MPa—the condition is performed.

Therefore, to achieve compressive strength at 7 and 28 days, respectively, $f_m^7 \geq 5$ MPa and $f_m^{28} = 15$ MPa, as well as a time availability of $\tau = 145$ minutes, the minimum cost of a binder is ensured when the low-clinker LSC is consumed at 266 kg/m³, LST consumption at 2 kg/m³, and lime-hydrated consumption at 10 kg/m³.

2.5.3 DRY MIXES FOR POROUS MASONRY MORTARS

To increase the thermal resistance of structures, the elimination of "cold bridges" in the constructions of buildings, it is advisable to use porous mortars. The effect of the use of such materials increases if the thermal conductivity of the masonry mortar is not more than that of the wall material. As is known, traditional cement-sand masonry mortars have high thermal conductivity (0.8–0.9 W/m·K), which is especially negatively reflected in the masonry from porous blocks, which they fasten [97].

To reduce the thermal conductivity of the mortars, light aggregates are introduced into their composition or pore-forming additives are used. The latter way, as a rule, is more expedient economically and technologically, considering the increased cost of light sand and difficulties with mixing mortars on light aggregates.

To create a porous structure, well-known porous additives for dry building mixes, for example, the UFAPORE system, can be used. It is advisable to make such mortars in the form of dry construction mixes ready to use, which in the required quantity can be quickly prepared directly at the construction site.

In the course of the research, mortars were prepared based on activated LSC, in which quartz sand with a maximum grain size of 1.25 mm was used as an aggregate. The mobility of the mortars at each point in the experimental plan with certain planning conditions (Table 2.26) was 7 to 9 cm.

The resulting mathematical models are given in Table 2.27, and the corresponding graphical dependencies in Figures 2.42 and 2.43.

When implementing the three-level plan B_3 in accordance with the planning conditions (Table 2.26), the W/C of the mortar mixture was determined, which provides the necessary mobility, average density, and compressive strength of samples that hardened in air-dry conditions at 7 and 28 days.

The analysis of the mathematical models of the strength of porous mortars (Figure 2.44), as expected, depends not primarily on the W/C, but on the porosity of the mortar, which can be estimated from the average density. With an increase in average density, the strength of the mortar increases regardless of the composition of the mortar.

A mixture based on activated LSC is characterized by an average density of mortars of 1300 to 1450 kg/m³ and a strength of 6 to 11 MPa. In this case, an increase in density from 1300 to 1450 kg/m³ is accompanied by an increase in strength by 50% to 60%. With an increase in the aggregate content (1:3), a linear decrease in strength at the same density is characteristic.

TABLE 2.26
Experiment Planning Conditions

Factors		Varying Settings			
Natural Value	Natural Value	−1	0	+1	Interval
Cement to aggregate ratio (C/Aggr.)	x_1	1:3	1:2	1:1	1
Additives UFAPORE content in mortar mix (kg/t)	x_2	0.25	0.5	0.75	0.25
Superplasticizer SP-1 consumption (kg/t)	x_3	0	1.5	3	1.5

TABLE 2.27
Experimental-Statistical Models of the Porous Mortars Based on Activated LSC

Parameters	Statistical Models
Water cement ratio	$W/C = 0.29 - 0.03x_1 - 0.04x_2 - 0.043x_3$ $- 0.001x_1^2 - 0.001x_2^2 + 0.044x_3^2 + 0.001x_1x_2$ $- 0.009x_1x_3 - 0.004x_2x_3$ (2.63)
Average density of mortar (kg/m³)	$\rho_0 = 1380 + 20x_1 - 62x_2 + 23x_3 + 3x_1^2$ $- 37x_2^2 - 2x_3^2 - 23x_1x_2 - 1x_1x_3 + 40x_2x_3$ (2.64)
Compressive strength of mortar at 7 days (MPa)	$f_m^7 = 3.65 + 2.58x_1 - 1.55x_2 + 0.98x_3 + 0.64x_1^2$ $- 0.025x_2^2 + 1.35x_3^2 - 1.36x_1x_2 + 0.37x_1x_3$ $+ 0.323x_2x_3$ (2.65)
Compressive strength of mortar at 28 days (MPa)	$f_m^{28} = 8.24 + 2.45x_1 - 1.37x_2 + 0.56x_3 - 1.9x_1^2$ $- 2.63x_2^2 + 1.038x_3^2 - 0.83x_1x_2 + 0.025x_1x_3$ $+ 0.68x_2x_3$ (2.66)

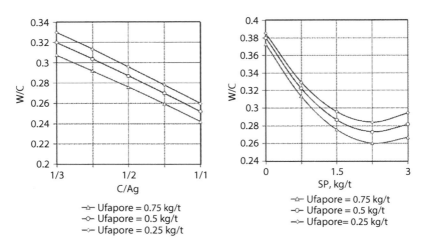

FIGURE 2.42 The influence of technological factors on the W/C of porous mortars.

2.5.4 DRY MIXES FOR SELF-LEVELING FLOORS

The current state of development of the construction industry and construction is characterized by an increase in interest and demand for dry mixes for the installation of Self-leveling floors. A feature of such floors is a quick set of strength, which allows them to be used immediately after installation. According to current standards, high demands are placed on the floors to provide a strength of 15 to 35 MPa, while the spreadability of the mortar mix should be at least 17 cm.

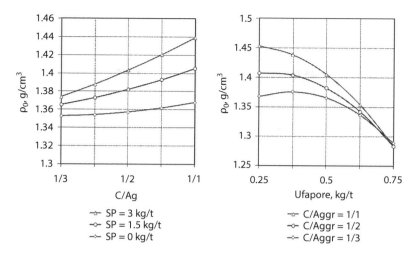

FIGURE 2.43 Influence of technological factors on the average density of porous mortars.

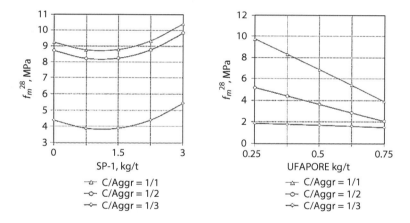

FIGURE 2.44 The influence of technological factors on the strength of porous mortars at 28 days.

To select the technological parameters to produce dry mixes for floors, experiments were carried out in accordance with the standard plan B_2 [40]. The experimental planning conditions are presented in Table 2.28.

The mobility of the mortars at each point of the plan was at least 17 cm. The statistical models of the W/C, as well as the compressive strength of the mortars at 3 and 28 days in the form of polynomial regression equations, are given in Table 2.29.

The corresponding graphical dependencies are shown in Figures 2.45 through 2.47.

With an increase in the hardening time, the compressive strength of Self-leveling flooring mortars based on activated LSC increases from 15.1 to 47 MPa, and the bending strength from 3.5 to 6.1 MPa; in this case, the W/C is about 0.40, and the

TABLE 2.28
Experimental Planning Conditions

Factors		Variation Levels			
Natural Value	Natural Value	−1	0	+1	Interval
Cement to aggregate ratio (C/Aggr.)	x_1	1:4	1:3	1:2	1
Additives superplasticizer Sika VC 225 content (kg/t)	x_2	0	1.2	2.4	1.2

TABLE 2.29
Experimental-Statistical Models of Mortars for Bulk Floors Based on Activated LSC

Parameters	Statistical Models	
Water cement (W/C) ratio	$W/C = 0.63 - 0.26x_1 - 0.17x_2$ $-0.003x_1^2 + 0.14x_2^2 - 0.02x_1x_2$	(2.67)
Compressive strength of mortar at 3 days (MPa)	$f_m^3 = 2.8 + 3.1x_1 + 2.7x_2$ $+ 2x_1^2 + 1.9x_2^2 + 1.1x_1x_2$	(2.68)
Compressive strength of mortar at 28 days (MPa)	$f_{m,tf}^{28} = 4.3 + 1.5x_1 + 0.3x_2$ $+ 0.3x_1^2 - 0.6x_2^2$	(2.69)
Compressive strength of mortar at 28 days (MPa)	$f_m^{28} = 17 + 11.3x_1 + 7.1x_2$ $+ 5.6x_1^2 - 4x_2^2 + 5.7x_1x_2$	(2.70)

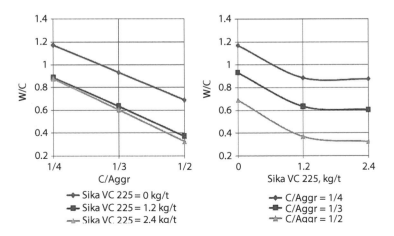

FIGURE 2.45 Influence of technological factors on the W/C of mortars for Self-leveling floors.

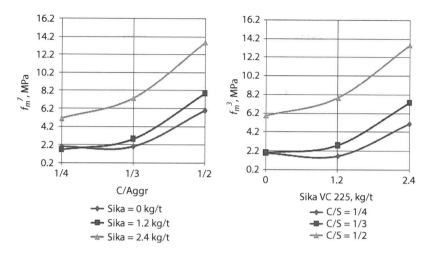

FIGURE 2.46 The influence of technological factors on the strength of mortars for Self-leveling floors 3 days.

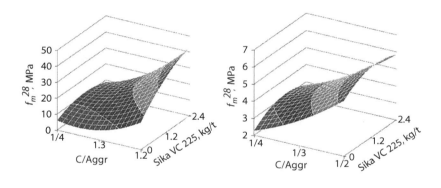

FIGURE 2.47 Influence of technological factors on the compressive and bending strength of mortars at 28 days.

spreadability of the mortar (when it flows from the Vicat cone) is in the range of 170 mm to 190 mm. The provision of such high strength indicators with a relatively high mobility of the mortar mixture is achieved by introducing Sika VC 225 polycarboxylate superplasticizer in dry mixes, an increase in the content of which to 2.4 kg/t leads to a decrease in W/C by about 45% to 50%.

3 Dry Construction Mixtures and Mortars Based on Them Using the Dust of Clinker-Burning Furnaces

3.1 CLINKER KILN DUST IS AN ACTIVE COMPONENT OF CEMENTITIOUS SYSTEMS

The Dust of a clinker-burning furnace includes particles of the fired raw material mixture for producing cement clinker with a grain size of less than 1 mm, which are carried out from the furnaces by exhaust gases and trapped in dust-precipitating devices. Emissions of dust generated during operation of furnaces make up 80% of the total amount of dust generated in cement production [98,99]. Depending on the size and type of furnaces, the presence of heat exchangers in them, the firing mode, the method of cement production, and the properties of the raw material mixture and fuel, dust removal on average ranges from 5% to 20% of the mass of the mixture fed to the furnace [98]. It was established [100] that the greatest amount of dust is carried out by flue gases at the end of the chain curtain zone, as well as the calcination zone. This dust is formed mainly as a result of mechanical destruction by chains and abrasion of raw granules on the lining. A noticeable amount of dust also can form in areas of exothermic reactions and cooling. Large dust removal is characteristic when using non-ductile raw materials, increasing the surface of heat exchangers and the gas flow rate.

In terms of material composition, the dust mainly consists of a mixture of undecomposed limestone and calcined clay [100]. Compared to the initial raw material mixture, dust is characterized by a higher content of carbonate component and a smaller content of clay and ferrous components. The dust composition may contain 10% to 20% of clinker minerals (β and γ C_2S, 8% to 10%, and C_2F and C_4AF, 10% to 12%). The content of free CaO increases as the content of clinker minerals in dust increases from 2% to 14%.

One of the characteristic features of dust is the increased content of alkaline compounds in it, represented by sulfates, carbonates, and bicarbonates of sodium and potassium, as well as silicates of variable composition. The alkali content in dust usually does not exceed 8% [101], although data [98] show that the total amount of $K_2O + Na_2O$ in dust captured by electrostatic precipitators can vary from 0.3% to 4.5% with the predominance of potassium oxide. The alkali content increases in the finest fractions. In the dust accumulated in the third field of the electrostatic precipitator it can be three to four times more than in the bunkers of the I and II fields.

The higher alkali content in dust is explained by the peculiarities of chemical processes occurring in rotary kilns during clinker roasting [102]. According to [102], as the temperature in the clinker kiln rises, alkalis bound in the raw mix and fuel ash mainly in the form of silicates are displaced by calcium oxide formed in the calcined material. A large number of free alkalis, starting from a temperature of about 800°C, diffuses to the surface of the calcined material and evaporates from it, part remains in the mixture and is part of the clinker, and part is converted into alkali metal carbonates, sulfates, and hydroxides, interacting with carbon dioxide and other components of the furnace gases. Alkalis remaining in the gas phase accumulate in the dust carried out from the furnace and captured by electrostatic precipitators.

Potassium, having a larger ionic radius, evaporates to a greater extent than sodium. Therefore, dust is enriched with potassium compounds. In the presence of SO_3 in the form of $CaSO_4$ in the feed mixture in the furnace gases and in clinker, low-volatile alkali metal sulfates are formed. Under the action of water vapor in the furnace gases and in clinker, alkali metal hydroxides are formed. The degree of alkali removal in the furnace gases is determined by the amount of dust removal and the temperature of the exhaust gases.

The dispersion of dust and its particle size distribution depends on the type of furnace units and dust collecting devices, as well as the characteristics of raw materials [103]. The dust carried with gases from rotary kilns is a polydisperse system, in which fine fractions up to 20 microns in size prevail. With the wet production method, the amount of fine fractions is from 40% to 70%, with the dry method up to 85% by weight. The number of fractions 0 to 10 microns and 10 to 20 microns in the dust of rotary kilns more than 100 m long, operating by the wet production method, is 18% to 50% and 10% to 20% by mass, kilns less than 100 m, 20% to 30% and 23% to 32%, and furnaces with cyclone heat exchangers 80% to 82% and 4% to 6%.

The dust from dust collecting chambers has a specific surface area of 2,500 cm^2/g to 3,200 cm^2/g, and from electrostatic precipitators 5,300 cm^2/g to 6,300 cm^2/g [104]. Increased dispersion is characteristic of dust formed during firing from a raw material mixture containing chalk and fine clay. When using denser and lower plastic materials, the dust contains larger particles.

The main way of utilizing the dust of clinker kiln rotary kilns is to return it to the production process of clinker burning. Various methods of dust return have been developed [99,105]: mixing it to sludge or raw flour entering the furnace in the form of dry powder, granules, or sludge; supply for chain curtains in a dusty or granular state; and blowing into the furnace from the hot end. A method also has been developed for firing dust after adjusting its composition in a separate furnace, which is a method for preliminary leaching of dust before returning it to the furnace.

Each of these methods of returning dust to the furnace has its advantages and disadvantages; however, returning dust in any way complicates the process of firing cement clinker and usually negatively affects the quality of cement due to an increase in the alkali content of the cement, an increase in the heterogeneity of the chemical composition, and deterioration of the structure of the clinker. The increased content of alkalis reduces the activity of Portland cement and causes its false setting [106], which leads to a rapid loss of plasticity of concrete and mortar mixtures and the need to increase water consumption, which degrades all the basic properties of concrete and mortars. With a high alkali content in cement, the risk of alkaline corrosion of concrete increases with the presence of reactive silica in the aggregates. In addition, excessive alkali content causes efflorescence [98,102].

Studies have established [99] that only with alkali oxides in the sludge of up to 0.7% to 0.8% can the entire amount of dust trapped by electrostatic precipitators be fed into the furnace without significantly affecting the quality of the clinker. Adding 5% to 15% of dust to the raw sludge can cause its coagulation and decrease in fluidity, sticking of the chain curtain, the formation of slurry rings, and a decrease in the resistance of the refractory lining of furnaces [107].

Along with the return of dust to the raw material mixture to produce clinker, a number of other areas of its application have been developed, including for the production of binders and mortars [108–113].

Dust generated during firing of Portland cement clinker and captured by electrostatic precipitators has weak astringent properties. They are due to the presence in the dust of a certain amount of clinker minerals formed as a result of predominantly solid-phase reactions. The interaction of free lime and the dehydrated clay contained in dust can make a specific contribution to the its hydraulic activity.

Dust activity can be in the range of 2.5 MPa to 7.5 MPa. Due to this activity, as well as high dispersion, which facilitates workability of mortar mixtures and prevents water separation, dust is recommended as an additive to mortars.

Along with masonry, plastering mortars are allowed, in which dust replaces lime. The cement-dust ratio depends on the composition and activity of the dust.

The hydraulic activity of dust increases significantly with the addition of gypsum. The addition of gypsum in an amount of 2% to 3% in the first stages of hardening increases the strength of samples by more than three times in comparison to samples from pure dust. At the same time, the increased content of gypsum negatively affects the final strength of the binder.

When dust is added to Portland cement, a decrease in strength is observed even with a dust content of more than 5% to 10% [98]. From experience, using the dust of clinker kilns as an additive produces lightweight grouting Portland cement [114].

The best results were obtained in the compositions from dust-blast-furnace granular slag and dust-slag-gypsum [115–117]. Along with the blast furnace slag, the possibility of using phosphorus and cupola slag in combination with dust also was noted. The theoretical prerequisite for the compatibility of slag and cement dust in the preparation of a composite binder is mainly the presence of soluble alkalis in the latter, which have the strongest activating effect on slag glass [31,118]. Dust components such as free oxide and calcium sulfate also have a certain activating effect on granular slag.

In the laboratory the Kiev Civil Engineering Institute, for the first time in a study of soil silicates, a slag-alkali binder based on blast-furnace granular slags and high-alkaline dust from clinker kiln electrostatic precipitators was obtained [115,117]. When determining the optimal ratio between slag and dust, it was found that the binder composition has the highest performance: 75% slag and 25% dust. When mixed with water, the 28 days compressive strength of binder varied from 20 MPa to 40 MPa depending on the amount of alkaline carbonates in the dust, and when mixed with alkaline mortars varied from 30 MPa to 70 MPa depending on the concentration of the alkaline solution.

In the products of hydration of a dust-slag binder, along with low-basic hydrosilicates of the tobermorite group, calcite, the complex salt formed by these compounds is identified. V. D. Glukhovsky and N. L. Macedon established [115–117] several peculiarities of dust-slag binders—an increase in activity proportional to the concentration of alkalis in dust increased bending strength. Based on these binders, it is possible to obtain mortars and concretes with a wide range of strength, frost resistance, and water tightness, having high adhesion to the reinforcement and reliably protecting it from corrosion.

Cementitious materials obtained by activating blast furnace or phosphorus slag cement dust are recommended for the construction of road bases from stone materials and soils [119,120].

It is recommended [120] to divide dust into three grades depending on indicators of chemical and hydraulic activity. A chemical indicator of dust activity is proposed to be calculated as the ratio of the total amount of CaO in the dust (CaO_{tot}) excluding CaO bound in gypsum, to CaO bound in dicalcium silicate, tricalcium aluminate, and tetracalcium aluminoferrite.

$$M = \frac{CaO_{tot} - 1.2LOI - 0.7(SO_3 - 0.85R_2O)}{2.8SiO_2 + 1.65Al_2O_3 + 0.7Fe_2O_3}, \quad (3.1)$$

where LOI is the loss on ignition.

The hydraulic indicator of dust activity (f) is determined by the formula:

$$f = f_{av}M, \quad (3.2)$$

where f_{av} is the average 28 days cement compressive strength ($f_{av} = 40$ MPa).

For dust of the first grade, the value of the module M should be more than 0.25, $f > 10$; of the second grade $M = 0.125$ to 0.25, $f = 5$ to 10; and of the third grade $M < 0.125, f < 5$.

Dust of the first grade can be used as an independent binder and dust of the second and third grades can be used as a component of dust and slag binder, respectively, in the amount of 15% to 25% and 20% to 30%.

3.2 RHEOLOGICAL AND STRUCTURAL-MECHANICAL PROPERTIES OF WATER PASTES OF DUST-SLAG-SUPERPLASTICIZER AND DUST-SLAG-CEMENT-SUPERPLASTICIZER SYSTEMS

The rheological properties of pastes formed by mixing binder with water to a large extent determine the manufacturability of mortar mixtures: their workability, preservation over time, compactibility, and stratification. Structural and mechanical properties of pastes significantly affect the processes of structure formation and hardening of mortars and concrete.

Normal consistency and setting time: At the first stage, the change in the normal consistency and setting time of pastes was studied herein in the dust-slag-superplasticizer (SP) and dust-slag-cement-SP systems. Composite binders were obtained in two ways:

1. Mixing ground blast furnace granulated slag (S_{sp} = 320 m²/kg), dust of clinker burning kiln, and superplasticizer SP-1 until obtaining a homogeneous mixture.
2. Joint grinding of these components to a certain specific surface.

Considering the inevitable fluctuations in the composition of the dust caused by fluctuations in the composition of the raw mixture and the operating parameters of firing, two dust samples of the cement plant (Dyckerhoff Ukraine) were selected for research, the chemical and grain composition of which, as well as the specific surface area, are given in Tables 3.1 and 3.2.

TABLE 3.1
The Chemical Composition of Dust

Dust Sample	Oxide Content (%)								
	CaO	CaO$_{free}$	SiO$_2$	Al$_2$O$_3$	Fe$_2$O$_3$	MgO	SO$_3$	R$_2$O	LOI
D_1	42.5	3.1	19.6	5.3	4.1	0.6	2.4	2.9	19.1
D_2	41.4	4.2	18.5	4.4	3.9	0.7	3.3	4.9	19.5

TABLE 3.2
Specific Surface and Grain Composition of Dust

Dust Sample	Specific Surface (m²/kg)	Content of Fractions in % by Weight with a Particle Size in Microns				
		0–20	20–40	40–60	60–90	90–200
D_1	396	51.5	25.4	9.3	11.5	2.1
D_2	410	53.3	23.6	10.2	10.3	2.5

TABLE 3.3
Normal Consistency and Setting Time of the Initial Components of Composite Binders

No.	Type of the Initial Components	Normal Consistency (%)	Setting Time (Hours-Minutes)	
			Initial	Final
1	Portland cement	25.4	2–5	4–10
2	Blast furnace slag	29.2	9–10	15–20
3	Dust fly (D_1)	40.5	4–30	9–40
4	Dust fly (D_2)	42.4	1–30	8–20

Pastes were prepared in a laboratory mixer with a constant duration of mixing the original binders with water. Normal consistency and setting time of the initial components of composite binders are given in Table 3.3. The results of determining the normal consistency and setting time of binders obtained by mixing the starting components are given in Table 3.4 and with joint grinding in Table 3.5.

The basic granulated slag used in the work, when grinding to a specific surface comparable with the specific surface of Portland cement, can set, although it is much slower than Portland cement (Table 3.1). The dust of clinker burning kilns sets much faster than blast furnace granulated slag, which can be explained by the content of a specific amount of clinker minerals in it, including tricalcium aluminate. An increase in the content of alkaline compounds in dust causes a significant acceleration in the initial of setting.

The combination of slag and dust allow significant acceleration of the setting time of compositions based on them, which indicates a significant intensification

TABLE 3.4
Normal Consistency and Setting Time of Pastes Based on Composite Binders Obtained by Mixing of the Initial Components

No.	Binder Composition (Ratio of Components by Weight)	Content SP (% by Weight)	Normal Consistency (%)	Setting Time (Hours-Minutes)	
				Initial	Final
1	Slag:dust (D_1)(1:1)	—	35	8–18	14–10
2	Slag:dust (D_2)(1:1)	—	33.4	6–15	12–20
3	Slag:dust (D_1)(1:1)	1	28.5	6–25	11–50
4	Slag:dust (D_2)(1:1)	1	28.2	5–15	10–40
5	Cement:slag:dust (D_1)(2:1:1)	—	31.5	3–30	6–50
6	Cement:slag:dust (D_2)(2:1:1)	—	27.4	2–40	5–30
7	Cement:slag:dust (D_1)(2:1:1)	1	25.5	2–30	5–10
8	Cement:slag:dust (D_2)(2:1:1)	1	24.8	2–10	4–30

TABLE 3.5
Normal Consistency and Setting Time of Pastes Based on Composite Binders Obtained by Joint Grinding of the Initial Components

No.	Type of Binder	The Content of the SP (% by Weight)	Specific Surface (m^2/kg)	Normal Consistency (%)	Setting Time (Hours-Minutes) Initial	Setting Time (Hours-Minutes) Final
1	Slag + dust (D_1)	—	350	29.2	2–10	8–30
2	Slag + dust (D_2)	1	370	23.1	1–40	6–40
3	Slag + dust (D_1)	—	580[a]	31.4	1–30	4–50
4	Slag + dust (D_2)	1	660[a]	21.8	0–40	1–30
5	Slag + dust (D_1)	—	340	28.4	1–40	6–30
6	Slag + dust (D_2)	1	380	22.6	1–10	5–50
7	Slag + dust (D_2)	—	590[a]	31.5	1–10	3–50
8	Slag + dust (D_2)	1	640[a]	22.9	0–10	1–10
9	Cement + slag + dust (D_1)	—	330	27.4	0–30	5–10
10	Cement + slag + dust (D_1)	1	360	22.3	0–38	4–30
11	Cement + slag + dust (D_1)	—	590[a]	29.5	0–25	1–10
12	Cement + slag + dust (D_1)	1	630[a]	22.5	0–21	0–40
13	Cement + slag + dust (D_2)	—	320	28.1	0–15	4–50
14	Cement + slag + dust (D_2)	1	350	22.7	0–20	3–39
15	Cement + slag + dust (D_2)	—	620[a]	31.4	0–17	0–45
16	Cement + slag + dust (D_2)	1	630[a]	23.1	0–15	0–37

Note: The composition of binder slag + dust is 1:1 (by weight), cement + slag + dust is 2:1:1 (by weight).

[a] Grinding intensifier—polyglycol (0.03% binder mass) was introduced.

of the processes of structure formation with a combination of these two substances. This effect can be explained by the combined activation of slag by alkalis, sulfates, and free lime contained in the dust.

A specific setting acceleration effect is observed (Tables 3.4 and 3.5) when superplasticizer SP-1 is introduced into the composition of the binder and the grinding fineness is increased. In the first case, the setting rate obviously is affected by a decrease in the water-binder ratio, in the second, by an increase in the reactivity of the binder, which contributes to the more rapid formation of a coagulation structure and crystalline nuclei of hydrated neoplasms.

The introduction of Portland cement during co-grinding to a binder based on blast furnace slag and dust from clinker burning kilns dramatically accelerates the setting time. For a three-component composite binder, obtained by co-grinding, setting times become much shorter than for Portland cement (initial of the setting time reaches 15 to 20 minutes (Table 3.5). With conventional mixing of Portland cement with ground slag and dust, the setting time of the binder is also accelerated, although to a much lesser extent (Table 3.4). Obviously, when the components are co-milled, mechano-chemical activation of the composite binder takes place.

Bringing in the binder the total amount of two-water gypsum to 7% to 8% (SO_3 content—4% to 5% by weight) allows adjustment of the setting time of composite binders to the recommended standard for Portland cement.

The normal consistency of composite binders fluctuates from 35% to 22%, depending on the composition, method of preparation, and dispersion. The binders obtained by co-grinding the components have a slightly lower normal consistency than when mixed. In the process of joint grinding, the passage of primary mechano-chemical processes is possible, and the amount of water demand significantly begins to deviate from that found from the additivity condition of the water demand of the individual components.

The most noticeable thinning effect on both two- and three-component composite binders is provided by the addition of superplasticizer. With a specific surface area of 320 m²/kg to 370 m²/kg, the introduction of an additive of 1% SP-1 by weight of the binder reduces its normal consistency by 18% to 21% (Figure 3.1). An increase in the fineness of grinding of binders to 580 m²/kg to 660 m²/kg causes a slight increase in normal consistency, compensated by the introduction of superplasticizer additives. A higher plasticizing effect of superplasticizers for binders with an increased specific surface is noted by other researchers [121].

The effect studied herein was the type and concentration of superplasticizer (SP) additives on the normal consistency of composite binders. Along with the influence of the additive SP-1 (SP of the naphthalene-formaldehyde type), the effect of the addition of SP polycarboxylate (Sika VC 225) type was determined. The experimental results are shown in Figure 3.1. These results show that the use of a polycarboxylate superplasticizer can achieve a greater decrease in the normal consistency of composite binders than naphthalene-formaldehyde, although the difference in the water-reducing effect with the introduction of both types of SP was not significant enough. In the first case, with the introduction of a 2% additive of SP from the weight of the binder, it is possible to reduce the normal consistency of three-component binders to 20% to 21%, and in the second case with 0.6% of the additive to 19% to 20%. The main water-reducing effect of the SP-1 additive in both two- and three-component binders is manifested at a content of 1%, Sika additives (0.3% by weight).

Viscosity and plastic strength: The viscosity of aqueous pastes based on composite binders largely determines the workability of mortar and concrete mixtures. To study the effect of the composition of pastes on their viscosity, the rotational rheometry method was used. The viscosity of the pastes was investigated using algorithmic experiments performed in accordance with the standard plan B_4 [40]. The viscosity of the pastes (η) was determined 2 to 3 minutes after kneading and calculated by the formula:

$$\eta = k \frac{P - P_0}{N} \tau, \qquad (3.3)$$

where k is the device constant, P is the cargo weight, N is the number of revolutions, τ is the duration of one revolution, and P_0 is the instrument idle load ($P_0 = 12.4$ г).

Dry Construction Mixtures and Mortars

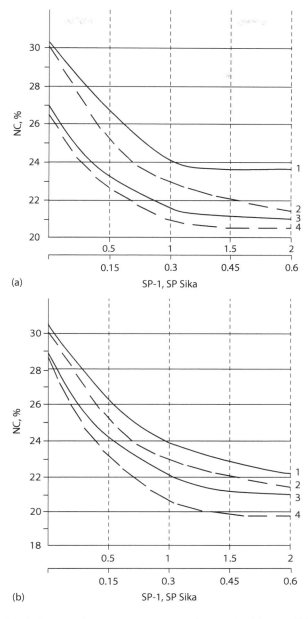

FIGURE 3.1 The influence of the type and amount of superplasticizer additives on the normal consistency of composite binders obtained by joint grinding of the initial components: — binders with the addition of SP-1%; - - - binders with Sp Sika. (a) $S_{sp} = 330$ m²/kg to 370 m²/kg; 1, 2: slag:dust, 1:1; 3, 4: Portland cement:slag:dust, 2:1:1, and (b) $S_{sp} = 580$ m²/kg to 660 m²/kg; 1, 2: slag:dust, 1:1; 3, 4: Portland cement:slag:dust, 2:1:1.

TABLE 3.6
Conditions for Planning Experiments When Studying the Viscosity of Pastes

No.	Factors Name	Coded View	Levels Varying -1	0	+1	Range of Variation
1	Volume concentration of dust in the dust-slag binder filler (DSF)—C_1	x_1	0	0.5	1	0.5
2	Volume concentration of DSF in the composite cement-dust slag binder—C_2	x_2	0	0.25	0.5	0.25
3	Volumetric concentration of composite binder in water paste—C_3	x_3	0.7	0.8	0.9	0.1
4	The content of superplasticizer (SP-1) in the composite binder (% by weight)	x_4	0	0.5	1	0.5

The conditions for planning experiments to determine the viscosity are given in Table 3.6, the matrix of the plan B_4, and the results of the experiments in Table 3.7.

The transition from volumetric concentrations to volumetric consumption of components in pastes in accordance with the experimental design conditions was carried out according to the formulas found from the condition:

$$C_1 + C_2 + C_3 = 1$$

or

$$\frac{V_d}{V_d + V_{sl}} + \frac{V_d + V_{sl}}{V_d + V_{sl} + V_c} + \frac{V_d + V_{sl} + V_c}{V_d + V_{sl} + V_c + V_w} = 1, \quad (3.4)$$

where V_d, V_{sl}, V_c, V_w is, accordingly, the specific volumetric consumptions per unit volume of the dust paste, blast furnace slag, cement, and water.

It follows from condition (3.4) that $V_d = C_1 C_2 C_3$; $V_{sl} = (1-C_1)C_2 C_3$; $V_c = (1-C_2)C_3$; $V_w = 1-C_3$.

Statistical processing of the experimental data made it possible to obtain the regression equations that are adequate at a 95% confidence probability, and can be considered as experimental statistical models of the effective viscosity of water pastes based on composite binders in the Portland cement—granulated blast furnace slag—clinker kiln dust system:

$$\eta_1 = 27.16 + 2.85x_1 + 5.48x_2 + 3.64x_3 - 5.26x_4 - 1.35x_1^2 + 1.99x_2^2$$
$$+ 0.49x_3^2 + 0.59x_4^2 + 3.74x_1 x_2 - 0.76x_1 x_3 - 0.36x_1 x_4 + 0.95x_2 x_3 \quad (3.5)$$
$$- 0.53x_2 x_4 - 0.33x_3 x_4$$

TABLE 3.7
Planning Matrix and the Results of Experiments to Study the Viscosity of Aqueous Pastes Based on Composite Binders

Plan Points	Factors				Viscosity (10^{-1} Pa·s)	
	x_1	x_2	x_3	x_4	η_1	η_2
1	+1	+1	+1	+1	38.5	32.3
2	+1	+1	+1	−1	53.2	49.4
3	+1	+1	−1	+1	31.4	25.3
4	+1	+1	−1	−1	42.6	39.3
5	+1	−1	+1	+1	19.7	18.7
6	+1	−1	+1	−1	27.8	28.1
7	+1	−1	−1	+1	16.3	15.4
8	+1	−1	−1	−1	25.6	26.1
9	−1	+1	+1	+1	28.1	35.4
10	−1	+1	+1	−1	38.5	41.4
11	−1	+1	−1	+1	19.5	16.1
12	−1	+1	−1	−1	27.8	27.5
13	−1	−1	+1	+1	23.5	24.3
14	−1	−1	+1	−1	33.3	32.8
15	−1	−1	−1	+1	15.8	15.5
16	−1	−1	−1	−1	24.8	25.5
17	+1	0	+1	0	29.4	28.3
18	−1	0	+1	0	22.3	21.6
19	0	+1	0	0	31.7	30.4
20	0	−1	0	0	26.7	27.8
21	0	0	+1	0	30.8	31.5
22	0	0	−1	0	24.6	23.7
23	0	0	0	+1	21.2	21.6
24	0	0	0	−1	34.4	35.5

Note: η_1 is the viscosity of the binder obtained by mixing the initial components; η_2 is the viscosity of the binder obtained by joint grinding of the initial components to $S_{sp} = 350$ to 380 m²/kg.

$$\eta_2 = 27.32 + 1.28x_1 + 4.64x_2 + 4.45x_3 - 5.66x_4 \\ - 2.41x_1^2 + 1.74x_2^2 + 0.24x_3^2 + 1.19x_4^2 + 2.23x_1x_2 - 1.68x_1x_3 \quad (3.6) \\ - 0.96x_1x_4 + 1.81x_2x_3 - 0.62x_2x_4 + 0.32x_3x_4$$

A graphical analysis of the main dependencies resulting from the obtained models is shown in Figure 3.2. It can be seen that in the studied factor space, the viscosity of pastes can vary by almost four times, reaching the highest values with a maximum dust content, a minimum water content, and the absence of superplasticizer additives. As the dust content in the dust slag filler decreases, its effect on the increase in viscosity decreases significantly. The content of superplasticizer in a paste based on composite binders is the most powerful compensation factor, stabilizing the increase in viscosity when the filler is saturated with dust. The second factor in the intensity of influence on the viscosity of pastes with a certain content of dust and slag filler is the volume concentration of the composite binder in the aqueous paste, which directly reflects the effect of the water-binder ratio.

Regression equations (3.5) and (3.6) allow, depending on the value of factor x_3 and, accordingly, on the water-binding ratio, regulation of the dose of superplasticizer, which provides constant viscosity depending on the filler content and the ratio of slag and dust in it (Figure 3.3). The nature of the influence of the studied factors on the viscosity is quite close for pastes obtained by mixing and co-milling of binder components.

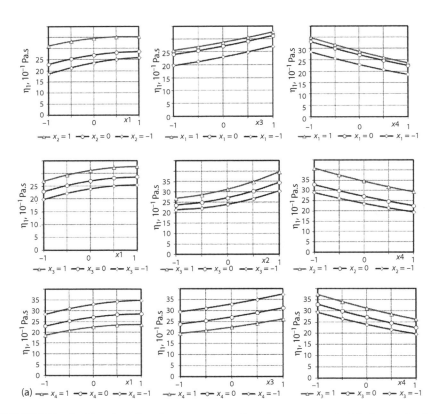

FIGURE 3.2 (a) The viscosity of aqueous pastes based on a composite binder obtained by mixing the initial components. *(Continued)*

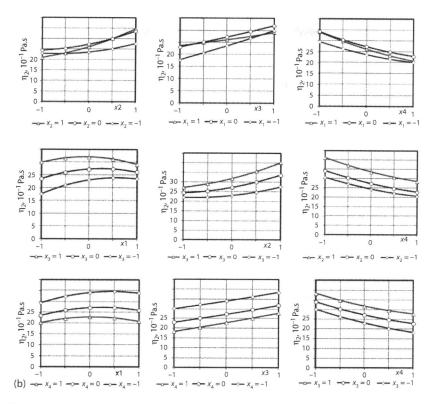

FIGURE 3.2 (Continued) (b) The viscosity of aqueous pastes based on a composite binder obtained by joint grinding of the starting components up to $S_{sp} = 350$ m²/kg to 380 m²/kg. (In addition to the two studies, the rest of the varied factors are at the zero (main) level.)

The characteristics of the initial hardening of viscous-aqueous dispersions are characterized by plastic strength, determined by the ultimate shear stress of the dispersed system in a certain period.

The plastic strength of the pastes was determined using a conic plastometer. The immersion depth of the cone was fixed by an indicator with a division price of 0.1 mm. The calculation of plastic strength was carried out according to the formula:

$$P_m = 0.096 \, P/h^2, \tag{3.7}$$

where P is the weight of the cone with the load (N) and h is the immersion depth of the cone in millimeters.

Plastograms for the characteristic compositions of pastes with water-binding ratios corresponding to their normal consistency are shown in Figures 3.3 through 3.5. A comparison of the duration of the first section of plastograms (Figures 3.4 through 3.6) with the duration of setting of pastes (Tables 3.4 and 3.5) shows their correspondence. The inflection points on the plastograms, separating the first and second sections,

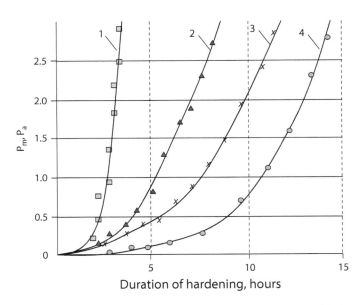

FIGURE 3.3 The plastic strength of pastes based on two-component binders composition slag-dust (D_1) 1:1 (by weight): (1) $S_{sp} = 660$ m²/kg, 1% SP-1; (2) $S_{sp} = 580$ m²/kg; (3) $S_{sp} = 370$ m²/kg, 1% SP-1; and (4) $S_{sp} = 350$ m²/kg.

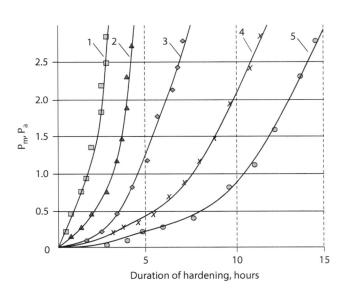

FIGURE 3.4 The plastic strength of pastes based on three-component binders Portland cement-slag-dust (D_1) 2:1:1 (by weight): (1) $S_{sp} = 630$ m²/kg, 1% SP-1; (2) $S_{sp} = 590$ m²/kg; (3) $S_{sp} = 630$ m²/kg, 1% SP-1, gypsum stone—3%; and (4) $S_{sp} = 360$ m²/kg, 1% SP-1; (5) $S_{sp} = 330$ m²/kg.

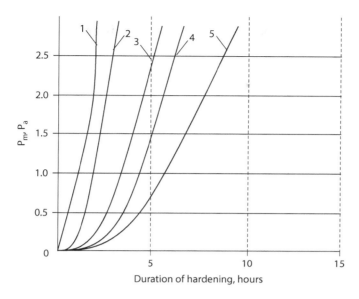

FIGURE 3.5 Plastic strength of pastes based on three-component binders Portland cement-slag-dust (D_2) 2:1:1 (by weight): (1) $S_{sp} = 630$ m²/kg, 1% SP-1; (2) $S_{sp} = 620$ m²/kg; (3) $S_{sp} = 630$ m²/kg, gypsum stone—3%; (4) $S_{sp} = 350$ m²/kg, SP-1%—1%; and (5) $S_{sp} = 320$ m²/kg.

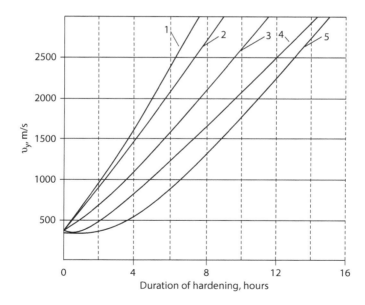

FIGURE 3.6 The ultrasonic waves velocity in hardening pastes based on composite binder cement: slag:dust (D_2) 2: 1: 1 (by weight). (1) $S_{sp} = 630$ m²/kg, 1% SP-1; (2) $S_{sp} = 590$ m²/kg; (3) $S_{sp} = 360$ m²/kg, 1% SP-1; (4) $S_{sp} = 330$ m²/kg; and (5) $S_{sp} = 630$ m²/kg, 1% SP-1, gypsum stone—3%.

approximately coincide with the final of the pasts setting. Factors contributing to the acceleration of setting time, a decrease in the water-binding ratio with the introduction of superplasticizer, and an increase in the specific surface also contribute to a reduction in the period of coagulation structure formation. These same factors contribute to the strengthening of the coagulation structure and the intensification of the formation of the crystallization structure. The pastes obtained based on three-component binders in the Portland cement-slag-dust system obtained by grinding to a specific surface of more than 500 m²/kg are characterized by a very short section on plastograms corresponding to coagulation structure formation. An additional introduction to pastes obtained based on binders with a high specific surface is two-water gypsum, which prolongs the period corresponding to the stability of the coagulation structure. At the same time, at the final setting, such pastes are characterized by a more intensive formation of a crystallization framework in the hardening structure.

Fluctuations in alkali content in dust, as indicated previously, lead to a certain acceleration of the elongation of setting; however, the final setting and, accordingly, the period of existence of the coagulation structure of pastes do not noticeably change.

The initial period of structure formation of pastes based on composite binders also was studied by an impulse ultrasonic method [122]. The transit time of ultrasonic waves was measured using an instrument at frequencies of 60 kHz with an accuracy of 0.25 μs to 0.5 μs. The measurement was carried out on freshly prepared cube samples with dimensions of 7.07 × 7.07 × 7.07 cm in plexiglas molds during hardening of the samples under normal conditions. The time, t_u, and the velocity of ultrasonic waves, v_u, were determined by the formulas:

$$t_u = f_2 - f_1 \tag{3.8}$$

$$v_u = l_b(f_2 - f_1) \tag{3.9}$$

where l_b is the base size of the sample ($l_b = 7.07$ cm) and $f_2 - f_1$ is the frequency difference.

On the curves of the speed of passage of ultrasonic waves, the initial period of formation of the coagulation structure is characterized by a horizontal section of Figure 3.6, the length of which is close in magnitude to the duration of the initial setting.

Determining the initial setting by the value of plastic strength is difficult due to its lower sensitivity to primary contacts between particles. As on the curves of changes in the plastic strength over time, the rate of formation of the crystallization structure can be related to the angle of inclination of the curves of the velocity of ultrasonic waves passing through hardening pastes. At 14 to 18 hours after mixing, the speed of passing of ultrasonic waves increases by four to five times. The value v_u as well as P_m is most significantly affected by the specific surface area of composite binders and the water-binder ratio.

Features of hydration of composite binders: The processes of hardening of composite binders in systems—blast furnace granular slag, dust, and Portland cement—are based on hydration of clinker minerals and slag glass under conditions of high alkaline oxide and SO_3 content.

Based on well-known theoretical concepts, the effect of the composition of pastes of two- and three-component composite binders on the degree of hydration and the kinetics of its change over time was studied herein [123–125].

Dry Construction Mixtures and Mortars

The *degree of hydration* of binders is one of the characteristic indicators of chemical activity, directly related to the strength of artificial stone based on them. For the first time a relationship was proposed [126] for the strength of cement stone ($f_{c.st.}$):

$$f_{c.st.} = AX^n, \tag{3.10}$$

where A is a constant characterizing the strength of the cement gel ($A \approx 240$ MPa), n is the cement characterization constant (for Portland cement $n \approx 2,6-3$), and X is the ratio of the volume of cement gel to the total volume of gel and voids (modified Feret structural criterion [40]).

According to [126], the parameter x can be found by the formula:

$$X = \frac{K_h V_c \alpha}{\alpha V_c + W/C} \approx \frac{0.647\alpha}{0.319\alpha + W/C}, \tag{3.11}$$

where $K_h = 2.09\text{–}2.2$ is the coefficient of increase in the volume of hydration products; V_c is the specific volume of cement ($V_c = 1/\rho_c = 0.319$ cm³/g); W/C is the water-cement ratio; α is the degree of hydration, that is, the part of cement that has passed hydration.

The criteria for the strength of cement stone, close to the criterion X [126], have been proposed by many researchers [127–129]. A. E. Sheikin linked the strength of the cement stone with the degree of hydration through the value of relative density (d_{rel}):

$$d_{rel} = \rho_{b.c.st}/\rho_{c.st} = (1+0.23\alpha\rho_c)(1+\rho_c W/C),$$
$$f_{c.st} = 310 d_{rel}^{2,7} \tag{3.12}$$

where $\rho_{b.c.st}$ is the average density of the cement stone in the dry state and $\rho_{c.st.}$ is the true density of the cement stone.

At the same time, as noted in [121], composite binders of low water demand are characterized by a lower degree of hydration, despite their high strength, both at an early age and during prolonged hardening. The discrepancy between the degree of hydration and strength is explained [130] by the composition and structure of hydrated neoplasms, as well as the structure of cement stone.

The degree of hydration was investigated by determining the amount of bound water [131]. From pastes of normal consistency, cubes were made with a 2 cm rib. They were stored under normal conditions. After a specified period of hardening, the samples were crushed into powder until they completely passed through a No.008 sieve. They were then treated with ethanol, dried at a temperature of 100°C to a constant mass, and calcined at 1000°C. The amount of chemically bound water was determined by the formula:

$$X = (a-b)/a, \tag{3.13}$$

where a is the mass of the dry sample before calcination and b is the mass of the sample after calcination.

The degree of hydration of binders α was calculated by the formula:

$$\alpha = x_1/x_2, \qquad (3.14)$$

where x_2 is the amount of bound water when fully hydrated.

The value of x_2 was determined after double heat-moisture treatment of samples with intermediate grinding.

The value of the degree of hydration of samples from pastes based on two- and three-component composite binders at 3, 7, and 28 days is given in Table 3.8.

An analysis of the data in Table 3.8 shows that the degree of hydration of two-component binders is 1.5 to 2 times lower than that of three-component ones. In both cases, the degree of hydration varies nonlinearly in time, the hydration rate decreases as the transition from 3 to 7 days, and then up to 28 days.

An increase in alkali content in dust leads to a slight increase in the degree of hydration of composite binders as a result of additional activation of the slag component. An increase in the degree of hydration by 20% to 40% can be achieved

TABLE 3.8
The Values of the Degree of Hydration of Pastes Based on Composite Binders Obtained by Co-Grinding the Starting Components

No.	Type of Binder (Ratio of Components by Weight)	The Content of the SP (% by Mass)	Specific Surface Area (m²/kg)	Degree of Hydration (Days)		
				3	7	28
1	Slag + dust (D_1)	—	350	0.20	0.31	0.35
2	Slag + dust (D_1)	1	370	0.17	0.28	0.32
3	Slag + dust (D_1)	—	580	0.24	0.36	0.40
4	Slag + dust (D_1)	1	660	0.21	0.34	0.39
5	Slag + dust (D_2)	—	340	0.22	0.38	0.43
6	Slag + dust (D_2)	1	380	0.20	0.31	0.37
7	Slag + dust (D_2)	—	590	0.25	0.37	0.41
8	Slag + dust (D_2)	1	640	0.21	0.37	0.42
9	Cement + slag + dust (D_1)	—	330	0.30	0.48	0.56
10	Cement + slag + dust (D_1)	1	360	0.27	0.46	0.55
11	Cement + slag + dust (D_1)	—	590	0.38	0.58	0.64
12	Cement + slag + dust (D_1)	1	630	0.34	0.55	0.62
13	Cement + slag + dust (D_2)	—	320	0.31	0.50	0.61
14	Cement + slag + dust (D_2)	1	350	0.29	0.49	0.60
15	Cement + slag + dust (D_2)	—	620	0.40	0.61	0.69
16	Cement + slag + dust (D_2)	1	630	0.36	0.59	0.67
17	Cement + slag + dust + gypsum stone 3% (D_1)	—	590	0.39	0.59	0.66
18	Cement + slag + dust + gypsum stone 3% (D_1)	1	630	0.35	0.55	0.63

Note: The composition of the binders slag + dust, 1:1 (by weight), cement + slag + dust, 2:1:1 (by weight).

Dry Construction Mixtures and Mortars

by increasing the specific surface area of binders from 350 m²/kg–380 m²/kg to 580 m²/kg–660 m²/kg. The content in the composition of the binder additive superplasticizer reduces the degree of hydration of composite binders, which is especially noticeable in the first stages of hardening.

In parallel with the determination of the degree of hydration, the strength and relative density of the cement stone also were determined on cubed samples with an edge of 2 cm. The results of the experiments are given in Table 3.9.

TABLE 3.9
Strength and Relative Density of Cement Based on Composite Binders

Type of Binder According to Table 3.8	Normal Consistency (%)	The Content of Super-plasticizer (% by Weight)	Specific Surface Area (m²/kg)	Strength (MPa) / Relative Density of Cement (Days)		
				3	7	28
1	24.2	—	350	8.3	14.2	18.4
				0.659	0.703	0.719
2	23.2	1	370	12.5	17.8	24.3
				0.659	0.704	0.720
3	31.4	—	580	11.2	16.3	21.5
				0.600	0.643	0.657
4	21.8	1	660	17.4	23.6	30.3
				0.692	0.746	0.767
5	28.4	—	340	8.8	15.6	19.5
				0.622	0.682	0.700
6	22.6	1	380	13.1	18.4	24.9
				0.678	0.723	0.748
7	31.5	—	590	12.4	17.6	25.2
				0.603	0.645	0.660
8	22.9	1	640	17.8	25.1	31.8
				0.679	0.744	0.765
9	27.4	—	330	19.2	26.8	35.4
				0.662	0.731	0.761
10	22.3	1	360	24.5	29.8	35.5
				0.711	0.789	0.827
11	29.5	—	590	23.4	28.5	37.8
				0.670	0.743	0.765
12	22.5	1	630	31.8	38.4	53.4
				0.737	0.824	0.852
13	28.1	—	320	21.5	28.2	35.8
				0.659	0.730	0.771
14	22.7	1	350	24.8	30.3	36.2
				0.714	0.796	0.841
15	31.4	—	620	24.5	31.6	38.3
				0.657	0.732	0.760
16	23.1	1	630	33.8	39.3	58.4
				0.737	0.831	0.864

From Table 3.9, the relationship between the strength and relative density of hardened composite binders is traced, which is consistent with the theoretical concepts proposed. A specific contribution to the value of the relative density is made by the degree of hydration and W/C, which corresponds to normal consistency, and the value of the latter becomes decisive for binders containing a superplasticizer with ordinary and with a high specific surface area.

3.3 TECHNOLOGICAL PARAMETERS OF OBTAINING COMPOSITE CEMENT-DUST-SLAG BINDERS OF LOW WATER DEMAND

The greatest effect with a significant content of mineral additives, is achieved, as is known [121], in the production of binders of low water demand, as well as finely ground multicomponent cements and concrete based on them. Upon receipt of these binders, the potential of the clinker component and mineral additives is realized to the maximum extent—their combined activating effect on the hydration and structure formation processes and synthesis of the properties of cement stone under conditions of minimum water content—and, as a result, minimum porosity is ensured.

The effect of the composition and specific surface of cement-dust-slag binders (CDSB) on their strength: The strength properties of binders of low water demand are determined by the content of clinker, the type and content of the mineral additive, superplasticizer introduced when grinding binders, and the fineness of grinding [121].

As shown previously, the addition of clinker kiln dust to Portland cement is advisable in compositions with blast furnace granulated slag, which not only eliminates the negative features of dust as an additive to cement (increased water demand, alkali, and free lime content), but also activates it under its influence.

To study the effect of the content and ratio of the main components, as well as the fineness of grinding on the strength indices of CDSB, experiments were carried out that were algorithmized in accordance with the standard plan B_4, and mathematical models of the compressive and bending strength of standard mortars at 2, 7, and 28 days were obtained.

The initial components in the implementation of experimental plan B_4 were Portland cement of the first type and dust (Tables 3.1 and 3.2) of clinker kilns (D_1), as well as blast-furnace granulated slag of the metallurgical plant. As a superplasticizer in the experiment used additives SP-1 and Sika VC 225. The grinding of binders was carried out in a laboratory mill. When grinding, the additive of superplasticizer was introduced in powder form.

The experimental planning conditions are given in Table 3.10. The experimental results obtained and the experimental statistical models are given in Tables 3.10 and 3.11.

Dry Construction Mixtures and Mortars

TABLE 3.10
Conditions for Planning Experiments in Studying the Effect of Composition and Specific Surface on the Strength of Composite Binders

Factors		Levels of Variation			
Natural	Coded	−1	0	+1	Interval of Variation
$(D_1 + Sl)$, %	x_1	10	35	60	25
$\dfrac{D_1}{D_1 + Sl}$	x_2	0	0.5	1	0.5
S_{sp}, m²/kg	x_3	300	450	600	150
SP, %	x_4	0	1	2	1

Note: D_1 is the dust of rotary kilns, Sl is the blast furnace granulated slag, S_{sp} is the specific surface, and SP is the superplasticizer SP-1.

TABLE 3.11
Experimental Results Obtained in the Study of the Effect of the Composition of CDSB on Their Strength Properties

No. of Points	x_1	x_2	x_3	x_4	Compressive Strength (MPa)			Bending Strength (MPa)		
					2 Days	7 Days	28 Days	2 Days	7 Days	28 Days
1	1	1	1	1	9.4	21.7	30.1	1.88	3.62	3.77
2	1	1	1	−1	3.9	14.6	21.6	0.79	2.43	2.7
3	1	1	−1	1	6.7	14.0	20.1	1.34	2.33	2.51
4	1	1	−1	−1	2.8	9.3	14.8	0.55	1.55	1.85
5	1	−1	1	1	15.4	40.7	58.6	3.08	6.79	7.32
6	1	−1	1	−1	10.9	33.5	47.8	2.19	5.58	5.97
7	1	−1	−1	1	12.2	29.9	44.3	2.43	4.99	5.53
8	1	−1	−1	−1	9.2	25.1	36.7	1.85	4.19	4.59
9	−1	1	1	1	22.5	47.0	66.9	4.49	7.84	8.37
10	−1	1	1	−1	17.5	39.9	58.4	3.51	6.65	7.3
11	−1	1	−1	1	16.8	37.3	53.1	3.35	6.22	6.64
12	−1	1	−1	−1	13.4	32.6	47.8	2.67	5.44	5.98
13	−1	−1	1	1	24.9	50.2	72.1	4.99	8.36	9.02
14	−1	−1	1	−1	21.0	42.9	61.3	4.2	7.16	7.67
15	−1	−1	−1	1	18.7	37.4	54.1	3.74	6.23	6.76
16	−1	−1	−1	−1	16.3	32.6	46.5	3.26	5.43	5.81
17	1	0	0	0	16.9	35.2	49.8	3.39	5.87	6.22
18	−1	0	0	0	27.0	51.6	73.1	5.40	8.6	9.13
19	0	1	0	0	15.7	25.6	36.1	3.15	4.27	4.52

(Continued)

TABLE 3.11 (Continued)
Experimental Results Obtained in the Study of the Effect of the Composition of CDSB on Their Strength Properties

No. of Points	x_1	x_2	x_3	x_4	Compressive Strength (MPa)			Bending Strength (MPa)		
					2 Days	7 Days	28 Days	2 Days	7 Days	28 Days
20	0	−1	0	0	20.2	35.1	49.7	4.04	5.85	6.21
21	0	0	1	0	19.3	39.1	55.1	3.86	6.51	6.89
22	0	0	−1	0	15.6	30.0	42.7	3.12	5.01	5.34
23	0	0	0	1	21.7	38.5	54.5	4.33	6.42	6.81
24	0	0	0	−1	17.8	32.6	46.4	3.55	5.43	5.8

The analysis of polynomial models (Table 3.11) obtained as a result of processing the experimental data allows tracing of the clearly expressed nonlinear nature of the influence of the studied factors on the strength of composite binders and finding their optimal values (Figures 3.7 and 3.8).

The influence of such factors as the content of dust in the mass of the binder and its specific surface is extreme. For the specific surface (Figures 3.7 and 3.8), it evident especially at 2 days and is in the range of 450 m²/kg to 500 m²/kg. Other authors [121,132,133] also note the indicated range of specific surface area as optimal for binders with low water demand. The introduction of a superplasticizer during grinding does not change the nature of the effect of the specific surface on strength (Table 3.12).

Some studies have shown that in the indicated range S_{sp} with favorable values of other factors and the content of SP-1 1% to 2%, binder normal consistence is 21% to 22% and 19% to 20% with the introduction of SP-1 and Sika VC 225, respectively. It is characteristic that the value of the normal consistence of binders with the addition of SP-1 does not increase significantly with increasing specific surface area. This conclusion can be explained by the fact that in cement with a higher specific surface, a specific increase in water demand is offset by a higher adsorption and, accordingly, water-retaining activity of the superplasticizer additive.

An increase in the content of dust and slag filler has a different effect on the strength of binders depending on the duration of hardening and the ratio of dust to slag (Figures 3.7 and 3.8). At an early age, an increase in the dust content of the binder almost linearly reduces its strength. To a lesser extent, strength reduced with the introduction of blast furnace slag and a combination of slag and dust. In this case, when combining slag and dust in a ratio of 1:1 by weight, the decrease in strength occurs approximately the same as when only one slag was introduced. At 28 days, an increase in the dust content in the binder from 10% to 30% reduces the compressive strength by 35%, and up to 50%—almost by half. A sharp drop in the strength of cement filled with dust is largely due to an increase in its water

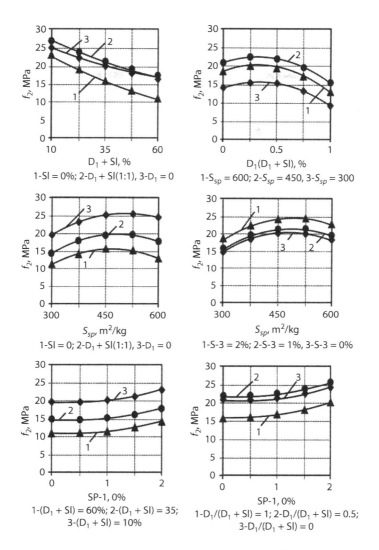

FIGURE 3.7 The dependence of the compressive strength of a composite cement-slag binder at 2 days on variable factors (Table 3.10). (The factors given in Table 3.10, the influence of which is not considered in Figure 3.7, are at the main (zero) level.)

demand (Table 3.13). An increase in the content of dust and slag filler in the ratio of dust:slag 1:1 by weight as well as one slag from 10% to 50% causes a decrease in strength by about 30% only (Figures 3.7 and 3.8). An increase in the content of dust in the dust slag filler from 0% to 50% does not cause an additional decrease in strength.

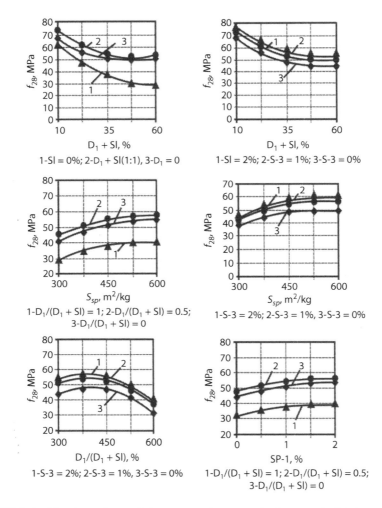

FIGURE 3.8 The dependence of the compressive strength of composite binder at 28 days from variable factors (Table 3.10). (The factors given in Table 3.10, the influence of which is not considered in Figure 3.8, are at the main (zero) level).

Reducing the water demand of binders due to the introduction of superplasticizer additives and increasing specific surface area can significantly increase their strength. A cement regrinding with a dust-fly content of 50% to 60% and the addition of SP-1 to a specific surface of 450 m^2/kg to 500 m^2/kg can increase the compressive strength from 30 MPa to 40 MPa. When the content of the complex additive containing clinker dust and blast furnace slag of 50% to 60% in a ratio of 1:1 by weight under the same conditions, it is possible to obtain a binder with a compressive strength of 50 MPa to 60 MPa in 28 days (Table 3.11; Figures 3.7 and 3.8).

TABLE 3.12
Mathematical Models of Strength Indicators

Strength Indicators		Mathematical Models	
Compressive strength (MPa)	2 days	$y_1 = 21.09 - 5.04x_1 - 2.24x_2 + 1.85x_3 + 1.96x_4$ $+ 0.88x_1^2 - 3.12x_2^2 - 3.62x_3^2 - 1.38x_4^2 - 0.88x_1x_2$ $- 0.75x_1x_3 + 0.13x_1x_4 - 0.13x_2x_3 + 0.25x_2x_4 + 0.38x_3x_4$	(3.15)
	7 days	$y_2 = 37.35 - 8.19x_1 - 4.74x_2 + 4.52x_3 + 2.98x_4$ $+ 6.04x_1^2 - 7x_2^2 - 2.8x_3^2 - 1.8x_4^2 - 3.97x_1x_2$ $- 0.5x_1x_3 + 0.78x_2x_3 - 0.03x_2x_4 + 0.61x_3x_4$	(3.16)
	28 days	$y_3 = 52.6 - 11.65x_1 - 6.78x_2 + 6.22x_3 + 4.03x_4$ $+ 8.82x_1^2 - 9.68x_2^2 - 3.68x_3^2 - 2.18x_4^2 - 5.81x_1x_2$ $- 0.94x_1x_3 - 1.06x_2x_3 - 0.56x_2x_4 + 0.81x_3x_4$	(3.17)
Bending strength (MPa)	2 days	$y_4 = 4.22 - 1.02x_1 - 0.45x_2 + 0.37x_3 + 0.39x_4$ $+ 0.17x_1^2 - 0.63x_2^2 - 0.73x_3^2 - 0.28x_4^2 - 0.18x_1x_2$ $- 0.15x_1x_3 + 0.03x_2x_4 - 0.03x_2x_3 + 0.05x_2x_4 + 0.08x_3x_4$	(3.18)
	7 days	$y_5 = 6.23 - 1.38x_1 - 0.79x_2 + 0.76x_3 + 0.5x_4$ $+ 0.99x_1^2 - 1.12x_2^2 - 0.48x_3^2 - 0.31x_4^2 - 0.66x_1x_2$ $- 0.08x_1x_3 + 0.01x_1x_4 - 0.13x_2x_3 + 0.005x_2x_4 + 0.1x_3x_4$	(3.19)
	28 days	$y_6 = 6.58 - 1.47x_1 - 0.85x_2 + 0.78x_3 + 0.51x_4$ $+ 1.09x_1^2 - 1.22x_2^2 - 0.47x_3^2 - 0.29x_4^2 - 0.73x_1x_2$ $- 0.12x_1x_3 + 0.01x_1x_4 - 0.13x_2x_3 + 0.07x_2x_4 + 0.1x_3x_4$	(3.20)

TABLE 3.13
Water Demand of Composite Binders and Mortars

No.	The Composition of Binder (%)				S_{sp} (m²/kg)	Normal Consistence %	Water-Binding Ratio Mortar
	Cement	Slag	Dust	Super-plasticizer SP-1			
1	65	—	35	—	310	32.3	0.41
2	40	—	60	—	310	36.5	0.43
3	65	—	35	1	320	28.3	0.40
4	65	—	35	2	320	27.8	0.40
5	65	35	—	—	310	27.1	0.40
6	40	60	—	—	310	28.2	0.40
7	40	60	—	1	325	22.2	0.37

(Continued)

TABLE 3.13 (Continued)
Water Demand of Composite Binders and Mortars

| | The Composition of Binder (%) | | | | | Normal | |
| | | | | Super-plasticizer | | Consistence | Water-Binding |
No.	Cement	Slag	Dust	SP-1	S_{sp} (m²/kg)	%	Ratio Mortar
8	40	60	—	2	325	21.4	0.36
9	50	25	25	—	320	29.1	0.41
10	50	25	25	1	330	22.7	0.36
11	50	25	25	2	330	21.6	0.35
12	50	25	25	1	460	21.7	0.35
13	50	25	25	1	580	22.1	0.36
14	50	25	25	0.6[a]	450	19.5	0.33
15	50	25	25	0.6[a]	590	19.8	0.33

Note: Gypsum and superplasticizer introduced into cement are considered in excess of 100%.

[a] Sika VC 225 polycarboxylate type superplasticizer was introduced into the binder of compositions No 14 and 15.

TABLE 3.14
Strength of Composite Binders with the Addition of Sika VC 225 Superplasticizer

| | The Composition of Binder (%) | | | | | Strength (MPa) | | | | | |
| | | | | | | Bending (Days) | | | Compressive (Days) | | |
No.	Cement	Slag	Dust	SP Sika	S_{sp} (m²/kg)	2	7	28	2	7	28
1	50	25	25	0.3	320	3.8	4.7	5.6	28	36	42
2	50	25	25	0.3	450	4.7	5.8	6.5	31	44	58
3	50	25	25	0.3	580	5.1	6.3	7.1	39	51	64
4	50	25	25	0.6	320	4.8	5.7	6.4	35	43	56
5	50	25	25	0.6	450	5.9	6.8	7.5	49	61	72
6	50	25	25	0.6	580	6.2	7.1	7.9	51	63	75
7	50	25	25	0.7	450	6.0	6.9	7.7	50	62	73

Note: Gypsum and superplasticizer introduced into cement are considered in excess of 100%.

Replacing naphthalene formaldehyde superplasticizer with polycarboxylate allows for a 50% complex dust and slag active filler to increase the 28-days compressive strength of the composite binder to 70 MPa (Table 3.14).

The resulting composite binders of low water demand are quick-hardening, their 2-day strength is not less than 50% at 28 days, and the ratio of compressive strength to bending strength for optimal compositions is within the usual range typical for Portland cement.

Considering the possibility of fluctuations in the content of alkaline compounds and free lime in dust, this study determined their effect on the strength indices of low-water demand CDSB containing dust and blast furnace granulated slag in the optimal ratio.

To this aim, additional experiments were carried out using dust (D_2) containing an increased amount of alkaline oxides (5.6%) and free lime (5.9%). The binder included 60% of dust and slag in a ratio of 1:1. The results of the experiments are given in Table 3.15.

As follows from the data given in Table 3.15, a slight increase in the content of R_2O and CaO_{free} in dust causes a slight tendency to increase the strength characteristics of composite binders with low water demand, which is associated with additional activation of the slag component.

The technological problem of obtaining binders with low water demand is the achievement of a high specific surface of the binder. In addition to the use of separator ball mills and other special types of mills for fine grinding in industrial conditions, it is of practical interest to use grinding intensifiers for this purpose. It was shown in [134] that the superplasticizer SP-1 can serve as a modifier of the properties of binder and concrete and an intensifier of grinding at the same time. With the same initial particle dispersion, the duration of clinker grinding with the optimal amount of gypsum and the addition of SP-1 to achieve a specific binder surface of 440 m²/kg was halved compared to the duration of clinker grinding without a modifier. It also was noted that the additive SP-1 facilitates the achievement of a high specific surface of the binder, preventing its aggregation during grinding. The physicochemical aspects of the processes occurring during the joint grinding of clinker and superplasticizer were considered in several works [121,135,136].

Using binders with clinker burning dust, it is rational to grinding it together with commercial Portland cement of the second type or slag Portland cement. A regrinding of three components together is also possible: Portland cement, ground slag, and dust. At present, industrial production lines have been developed for obtaining such binders by grinding cement and other mineral components to the

TABLE 3.15
The Effect of Alkali and Free Lime in the Dust of Clinker Kilns on the Strength of CDSB

Dust Sample	Content SP	Specific Surface Area (m²/kg)	Strength (MPa)					
			Bending (Days)			Compressive (Days)		
			2	7	28	2	7	28
D_1-1 (R_2O = 2.9, CaO_{free} = 3.1%)	SP-1—1%	450	3.4	5.9	6.2	16.9	35.2	49.8
	SP-1—2%	450	3.5	6.1	6.5	17.6	36.4	51.6
	Sika-0.6%	450	5.9	6.8	7.5	49.0	61.0	72.0
D_1-2 (R_2O = 5.6, CaO_{free} = 5.9%)	SP-1—1%	450	3.6	6.5	6.9	18.6	39.4	53.3
	SP-1—2%	450	3.8	6.7	7.1	19.6	39.7	55.7
	Sika-0.6%	450	6.3	7.2	7.8	51.0	63.0	74.0

required specific surface using ball and vibration mills. The latter create better conditions for achieving a high specific surface of binders and their mechano-chemical activation [136].

With fine grinding of powders in both ball and vibration mills, it is important to prevent the accumulation of small particles on the grinding bogies and aggregation of particles.

The kinetics of grinding composite binders with ball and vibration mills was studied in the laboratory using Portland cement of the first type (S_{sp} = 310 m²/kg), ground blast furnace slag (S_{sp} = 240 m²/kg), dust of clinker kilns (S_{sp} = 390 m²/kg). During grinding, SP-1, Sika, and Propylene glycol additives were added. The experimental results are shown in Figures 3.9 and 3.10. It follows from these results that when grinding in a laboratory ball mill after reaching a specific surface of about 400 m²/kg, a further increase in the grinding time leads to a slight decrease in S_{sp} due

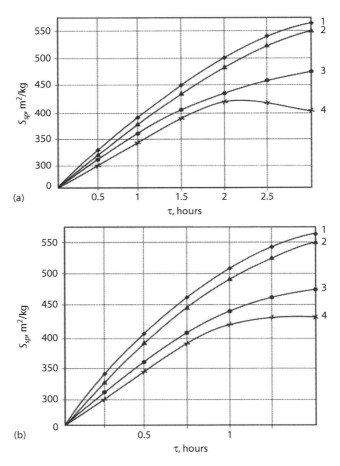

FIGURE 3.9 The grinding kinetics of CDSB with additives Sika and propylene glycol: (a) in a ball mill, (b) in a vibratory mill: (1) 0.04% propylene glycol+0.6% Sika, (2) 0.04% propylene glycol, (3) 0.6% Sika, and (4) no additives.

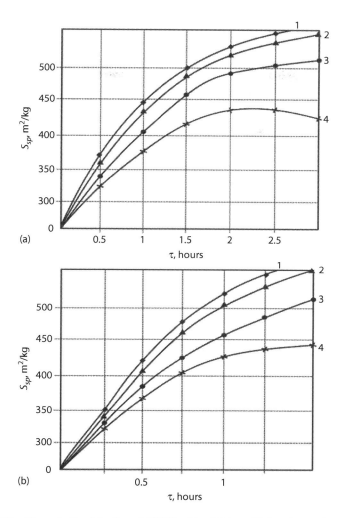

FIGURE 3.10 The grinding kinetics of CDSB with additives SP-1 and propylene glycol: (a) in a ball mill, (b) in a vibration mill: (1) 0.04% propylene glycol+1% SP-1, (2) 0.04% propylene glycol, (3) 1% SP-1, and (4) no additives.

to aggregation of fine particles. The addition of SP-1 intensifies the grinding, making it possible to reach S_{sp} = 500 m²/kg. Moreover, as the grinding time increases, the intensification effect of the grinding is more noticeable, although after 2 hours of grinding, the growth of the specific surface of the binder slows down. The intensification of grinding with the introduction of propylene glycol, as well as the composition of propylene glycol and SP-1, is even more noticeable. As is known, propylene glycols, which are representatives of the glycol group, like surfactants of the amine group, are strong grinding intensifiers in the concentration range of 0.03% to 0.1% by weight of the binder.

TABLE 3.16
The Effect of Propylene Glycol Additives on the Strength Characteristics of CDSB

		Content of Additives Superplasticizer (%)	Content of Propylene Glycol Additives (%)	Strength (MPa)					
				Bending (Days)			Compressive (Days)		
No.	S_{sp} (m²/kg)			2	7	28	2	7	28
1	430–450	SP-1—1.5	—	4.8	5.7	6.6	33	54	61
			0.03	4.4	5.9	6.3	34	51	63
			0.04	4.3	5.1	6.5	32	51	61
			0.05	3.6	4.4	5.9	28	46	59
2	430–450	Sika—0.6	—	5.7	6.5	7.3	47	59	71
			0.03	5.7	6.1	7.5	46	58	72
			0.04	5.5	6.1	7.1	45	57	70
			0.05	4.6	5.5	6.8	39	55	66
3	430–450	—	—	3.4	4.1	5.4	25	37	45
			0.03	3.5	3.9	5.2	24	35	44
			0.04	3.2	3.7	5.1	23	35	43
			0.05	2.8	3.2	4.6	19	32	41

Note: The composition of the binder: Portland cement of the first type M-500, 50%; blast furnace slag, 25%; fly ash, 25%.

Grinding in a vibratory mill made it possible to obtain binders with a specific surface area like that achieved in a ball mill about two times faster. The introduction of additives SP-1 and propylene glycol during grinding in a vibratory mill as well as in a ball mill leads to a significant intensifying effect.

The effect of superplasticizer Sika as a grinding intensifier was less than SP-1 during grinding in a ball mill and in a vibratory mill. The increase in the binder content of these additives more, respectively, 1% and 0.6% did not entail a marked increase in their action as intensifiers of grinding.

As follows from the obtained experimental data, in both cases, with a close specific surface, binders with propylene glycol additive in the range of 0.03% to 0.04% by mass have practically the same strength as without this additive (Table 3.16). A further increase in the amount of propylene glycol leads to a decrease in strength, especially in the early stages of hardening, which can be explained by the adsorption effect of stabilization of hydration processes [137,138].

3.4 DRY MIXES AND MORTARS BASED ON DUST SLAG BINDERS

The properties of dry construction mixtures and mortars intended for masonry work were investigated using composite dust binders (CDSB), containing additionally blast furnace granulated slag and Portland cement.

In the practice of modern construction, cement-lime mortars have found the greatest application for masonry work [40]. Topical for construction from the standpoint

Dry Construction Mixtures and Mortars

of resource and energy intensity of mortar and masonry work are the problems of reducing cement consumption and the rejection of the use of lime due to the use of surfactant additives and technogenic mineral raw materials.

The influence of the composition of dry mixtures containing dust, on the properties of mortars: Dry mortar mixtures were prepared by thorough mixing in a laboratory mixer of Portland cement, fly ash, ground blast furnace granulated slag, and sand with the introduction of a powdered additive of superplasticizer SP-1. The slag was ground to a specific surface S_{sp} = 310 m²/kg to 330 m²/kg in a laboratory ball mill. The dust corresponded to the chemical composition given in Table 3.1. To achieve the necessary uniformity of the mixtures, they were mixed in two stages: at the first stage, the components of the binder, at the second, with the introduction of sand.

To determine the effect of the composition of dry mixtures on the normalized properties of mortars, experiments were carried out algorithmized in accordance with a semi-replica 2^{4-1} of a full-factor experiment with four factors varying [139]. The experimental planning conditions are given in Table 3.17.

The consumption of quartz sand at each point of the matrix was determined from the condition: S = 1000-C-Sl-D, where C, Sl, D is the volume consumption of cement, slag, and dust in liters per cubic meter.

The sand used was fine with a maximum grain size of 1.25 mm (M_f = 1.3). Mortar mixtures at each point of the matrix were obtained by mixing the dry mixture while adjusting to mobility 8 to 10 cm by immersion of a standard cone.

The output parameters were determined by the compressive strength of mortars for samples hardening in air-dry conditions, the adhesion to the base, and frost resistance. The compositions of the mixtures and the results of experiments to determine the properties of mortars based on them are given in Table 3.18.

Additionally, the output coefficient of the mortar mixture (L/kg) and the binder-water ratio necessary to achieve the given mobility of the mortar mixture also were found. As a result of statistical processing of the experimental data, incomplete quadratic regression equations of the studied parameters from varying factors are obtained, which are given in the coded variables in Table 3.19. Along with the linear effects of the influence of individual factors, equations (3.21) through (3.26) consider the effects of their pairwise interactions.

TABLE 3.17
Conditions for Planning Experiments When Studying the Effect of the Composition of Dry Mixtures on the Properties of Mortars

Factors		Varying Settings			
Natural	Coded	−1	0	+1	Interval
Cement consumption (C) (kg/t)	x_1	100	150	200	50
Slag consumption (Sl) (kg/t)	x_2	50	75	100	25
Dust consumption (D_1) (kg/t)	x_3	50	75	100	25
Superplasticizer (SP-1) consumption (% by weight of binder)	x_4	0	0.5	1	0.5

TABLE 3.18
The Compositions of Dry Mixtures and the Properties of Mortars Based on Them

No.	Component Consumption (kg/t)				The Output Coefficient of the Mortar (K_b), l/kg	Binder-Water Ratio (B/W)	Compressive Strength (MPa)		Adhesion to Base, MPa (f_{adh})	Frost Resistance, Cycles (F)
	Cement (C)	Slag (Sl)	Dust (D_1)	Sand (S)			7 Days (f_m^7)	28 Days (f_m^{28})		
1[a]	200	100	100	600	0.7	2.94	21.8	27.6	0.82	100
2	200	100	50	650	0.5	2.33	17.4	23.5	0.61	75
3	200	50	100	650	0.5	2.27	8.5	14.2	0.45	50
4[a]	200	50	50	700	0.65	2.63	21.0	30.0	0.74	75
5	100	100	100	700	0.6	1.92	7.1	13.5	0.43	35
6[a]	100	100	50	750	0.65	2.08	6.7	13.8	0.34	50
7[a]	100	50	100	750	0.65	2.5	8.1	15.4	0.51	50
8	100	50	50	800	0.55	1.35	3.4	7.4	0.25	25

[a] In the mixture of compositions 1, 4, 6, and 7, the superplasticizer SP-1 was introduced in an amount of 4, 3, 2.5, and 2.5 kg/t, respectively.

Dry Construction Mixtures and Mortars

TABLE 3.19
Mathematical Models of the Properties of Mortars Based on Dry Mixes

Parameters	Statistical Models	
Compressive strength of mortar at 7 days (MPa)	$f_m^7 = 11.8 + 5.43x_1 + 1.5x_2 - 0.38x_3 + 2.65x_4$ $+ 0.93x_1x_2 - 1.65x_1x_3 + 1.58x_1x_4$ $+ 1.58x_2x_3 - 1.65x_2x_4 + 0.92x_3x_4$	(3.21)
Compressive strength of mortar at 28 days (MPa)	$f_m^{28} = 18.0 + 5.81x_1 + 1.59x_2 - 0.3x_3$ $+ 3.69x_4 + 0.14x_1x_2 - 2.6x_1x_3 + 1.29x_1x_4$ $+ 1.29x_2x_3 - 2.6x_2x_4 + 0.14x_3x_4$	(3.22)
Adhesion to base (MPa)	$f_{adh} = 0.52 + 0.14x_1 + 0.03x_2 - 0.03x_3$ $+ 0.08x_4 + 0.03x_1x_2 - 0.05x_1x_3 + 0.04x_1x_4$ $+ 0.04x_2x_3 - 0.5x_2x_4 + 0.14x_3x_4$	(3.23)
Frost resistance (Cycles)	$F = 57.5 + 17.5x_1 + 7.5x_2 + 1.25x_3$ $+ 11.25x_4 + 5x_1x_2 - 1.25x_1x_3 + 1.25x_1x_4$ $+ 1.3x_2x_3 - 1.3x_2x_4 + 5x_3x_4$	(3.24)
Binder-water ratio	$B/W = 2.25 + 0.29x_1 + 0.06x_2 + 0.15x_3$ $+ 0.28x_4 + 0.028x_1x_2 - 0.09x_1x_3 - 0.04x_1x_4$ $- 0.04x_2x_3 - 0.09x_2x_4 + 0.028x_3x_4$	(3.25)
Output coefficient of the mortar (K_o)	$Ko = 0.606 + 0.06x_1 + 0.031x_2 + 0.006x_3$ $+ 0.031x_4 + 0.006x_1x_2 + 0.006x_1x_3 + 0.007x_1x_4$ $+ 0.005x_2x_3 + 0.0007x_2x_4 + 0.006x_3x_4$	(3.26)

An analysis of the equations describing the influence of compositional factors of dry mixtures on the compressive strength of mortars shows (Figures 3.11 and 3.12) that in the studied range of their changes, it is possible to obtain mortars with a 28-day compressive strength of 10 MPa to 30 MPa. By 7 days, the strength of the mortars reaches 70% of the 28-day strength. Strength naturally decreases with a change in the composition of the mixtures, leading to increase in the binder-water ratio. In this regard, the effect of cement consumption and the addition of superplasticizer is especially evident.

In the regression equations of 7- and 28-day strength, the most significant linear effects are manifested for factors x_1 and x_4, that is, the consumption of cement and superplasticizer. An increase in the content of slag in the composition of the mixture can cause either a slight increase or a decrease in the 28-day strength of the mortars. An increase in the strength of mortars with an increase in the slag content is characteristic with a reduced cement content (Figures 3.11 and 3.12). Conversely, in mixtures containing an increased cement consumption, a decrease in the strength is observed with an increase in the slag content. Reducing the dust content in the

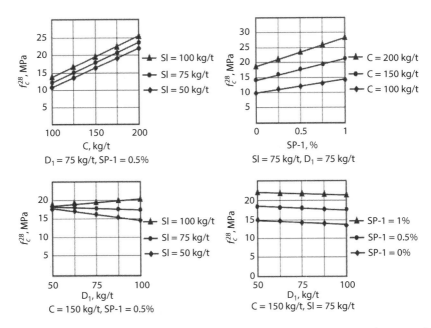

FIGURE 3.11 Influence of the dry mixture composition on mortars compressive strength at 28 days (SP, superplasticizer; D, dust; C, cement; and Sl, slag).

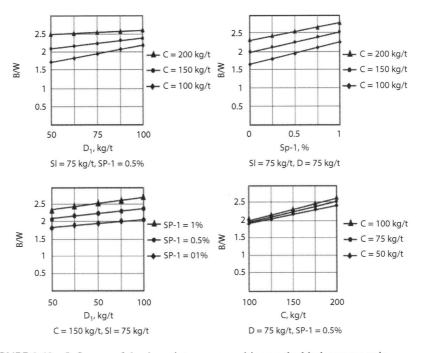

FIGURE 3.12 Influence of the dry mixture composition on the binder-water ratio.

Dry Construction Mixtures and Mortars

selected range also can reduce or hardly affect or increase the strength. A slight increase in the strength of the mortars with an increase in the dust content is noted with a simultaneous increase in the content of slag. With insufficient slag content in mortar mixtures, an increase in dust content leads to a decrease in strength. Strength reduction is also possible in the absence of superplasticizer in the mortar mixture (Figure 3.11).

In relatively low-cement mortars, the combined introduction of slag and dust as active mineral fillers in an amount of up to 200 kg/m³ each leads to a noticeable increase in 7- and 28-day compressive strength.

The effect of the dry mixture composition on the binder-water ratio and the mortars output coefficient are shown in Figures 3.12 and 3.13.

Experimental-statistical models can be used for designing the compositions of dry mixtures and the mortars based on them. The composition of mixtures providing a given strength is easy to select by using equations (3.21) and (3.22) and considering possible resource limitations. For this purpose, it is possible to use a nomogram (Figure 3.14).

To provide a set of necessary properties, for example, 28-day compressive strength, adhesion to the base, and frost resistance, a system of three corresponding equations is solved (the consumption of superplasticizer x_4 is set) and the necessary consumptions (in kg/t) of cement, slag, and dust (x_1, x_2, and x_3) are obtained.

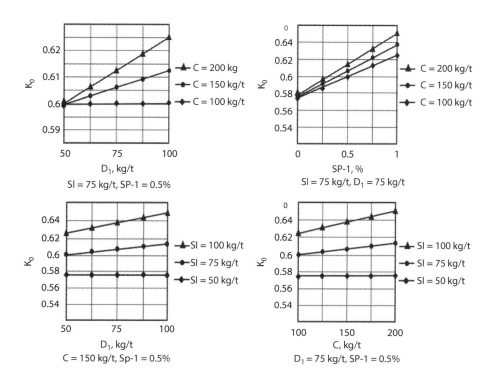

FIGURE 3.13 Influence of the dry mixture composition on the mortar output coefficient.

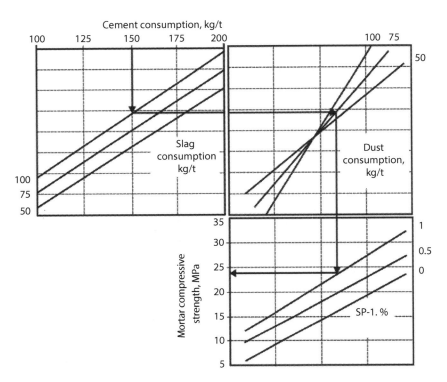

FIGURE 3.14 Nomogram for calculating of the dry mortar composition to obtain a masonry mortar.

In the transition from the composition of the dry mixture (kg/t) to the composition of the mortar by calculation using the appropriate regression equation (3.27), the output coefficient and calculate the consumptions of each component is found according to the formula:

$$CS_{k.m.} = \frac{CS_{d.m}}{K_0}, \quad (3.27)$$

where $CS_{k.m.}$ is the consumption of components of the mortar mixture in kilograms per cubic meter, $CS_{d.m}$ is the consumption of the component of the dry mixture in kilograms per ton, and K_0 is the output coefficient in cubic meters per ton.

If necessary, a cost analysis of the found compounds is possible using the condition:

$$\text{Cost}_m = C_c C + C_{sl} SL + C_d D + C_s S + C_{sp} Sp, \quad (3.28)$$

where Cost_m is the total cost of materials per 1 ton of dry mix or 1 m³ of mortar mixture, C_c, C_{sl}, C_d, C_s, C_{sp} is the cost of cement, blast furnace slag, dust, sand, and superplasticizer, and C, Sl, D, S, Sp is the specific consumption of the respective components.

Dry Construction Mixtures and Mortars

TABLE 3.20
The Calculated Compositions of Dry Mixtures and Mortars

Compressive Strength of Mortars at 28 Days (MPa)	Component Consumption (kg/t)					Output Coefficient	Binder-Water Ratio
	Cement	Slag	Dust	Sand	Superplasticizer (%)		
5	100	75	50	775	0	0.53	1.43
7.5	100	75	50	775	0.3	0.54	1.60
10	200	75	50	675	0.5	0.65	2.48
150	200	50	100	650	1	0.67	2.91
20	105	100	100	695	1	0.62	2.47
25	140	100	50	710	1	0.64	2.29

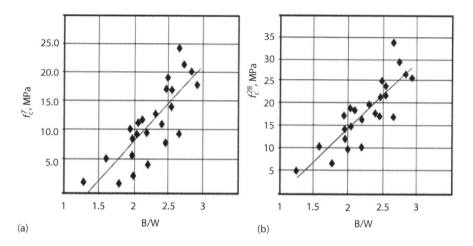

FIGURE 3.15 The relationship of the strength of mortars with a binder-water ratio: (a) 7 days and (b) 28 days.

Table 3.20 shows the compositions of dry mixtures of various grades in strength, calculated using equations, ensuring the lowest possible cement consumption considering the range of variation of factors adopted in Table 3.17.

Figure 3.15 shows the calculated dependences characterizing the relationship between the strength of mortars and the binder-water ratio.

The dependence of the strength of the mortar on the binder-water ratio (B/W) can be interpreted by formulas of the type:

$$f_c = K(\text{B/W} - b), \tag{3.29}$$

where K and b is the coefficients that consider the features of the materials used. For materials used at 7 days, $K = 12.7$, $b = 1.32$, and at 28 days, $K = 15$, $b = 1.05$.

TABLE 3.21
The Calculated and Experimental Values of the Strength of Mortars

Composition Number	Calculated Values of 28 Day Compressive Strength (MPa)	Experimental Values of 28 Day Compressive Strength (MPa)	Relative Error (%)
1	5.7	6.3	10
2	8.3	7.7	15
3	21.5	24.1	12
4	27.9	25.4	9
5	21.3	23.6	11
6	18.6	16.7	11

The coefficient K should reflect to a certain extent also the influence of the activity of the used binders. When considering the influence of the binder's compressive strength (R_b) formula (3.29) will look like:

$$f_c = AR_b(B/W - b). \tag{3.30}$$

The calculated and experimental values of the strength of the mortars calculated by the formula (3.30) are given in Table 3.21.

An important indicator of the quality of masonry mortars is the adhesion to the base, measured by the avulsion strength, determined by their cohesive and adhesive properties. The main cohesive parameter of a mortar is its compressive strength, discussed previously. The adhesion of cement mortars to the bases is a complex dependency of many factors, including the dispersion of the binder, the wettability of the base with a mortar mixture, etc. [121].

The main role in the formation of adhesive strength is assigned to the ratio of the surface energy of the adhesive and the substrate. The value of surface energy is component of the total energy of substances, determined by the characteristics of their structure, chemical composition, dispersion, etc.

The wetting of the base with a mortar and an increase in adhesive strength are facilitated by a decrease in surface energy when creating an adsorption-active medium due to surfactant additives [138].

The influence of the adsorption-active medium increases with increasing dispersion of the filler and its concentration, which is associated with an increase in the interphase surface and, accordingly, with excess surface energy.

The adhesion strength of mortar on composite binders, including the dust of clinker kilns and blast furnace granulated slag for all compositions in the studied area of factor changes exceeds the regulatory requirements. It increases significantly with an increase in the volume of the binder in dry mixes (Figure 3.16). The decrease in the water-binding ratio achieved by the addition of a superplasticizer also has a positive effect, which, obviously, should improve the cohesive properties of the mortar. The addition of a superplasticizer related to hydrophilic surfactants should help to improve the wetting of the surface of the base and increase the adhesive strength.

Dry Construction Mixtures and Mortars

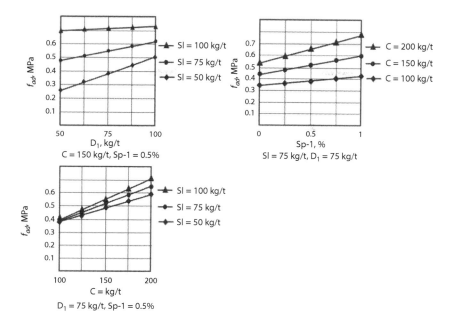

FIGURE 3.16 The effect of the composition of the dry mixture on the adhesion strength of mortars with the base (C, Sl, D_1, SP: consumptions of cement, slag, dust, superplasticizer respectively).

The addition of clinker dust into the mortar mixture positively affects the adhesion strength. The degree of its influence with an increase in the specific content in the mortar mixture increases with a simultaneous increase in the content of slag and a decrease in the cement content (Figure 3.16). This result can be explained by the improvement in the astringent properties of the dust-slag composition when the optimum ratio of components is achieved. At the same time, the effect of a cement-dust slag binder on the adhesion of the mortar to the base becomes especially significant with a lack of cement paste (Figure 3.16). The positive effect of dust as a highly dispersed component of the dry mixture, increasing its water demand, on the adhesion strength increases with the addition of superplasticizer additives, which reduces the negative effect of dust on water content. An increase in the positive effect of dust of clinker kiln on the adhesion strength of the mortar to the base is indicated by the presence in the corresponding regression equation (Table 3.19) of the most significant effect of the interaction of factors x_3 and x_4. The positive effect of dust on the adhesive properties of mortar mixtures is also facilitated by its chemical composition. The greatest adhesion strength are mixtures consisting of cement and limestone (limestone grains prevail in dust), the smallest is a mixture of quartz sand.

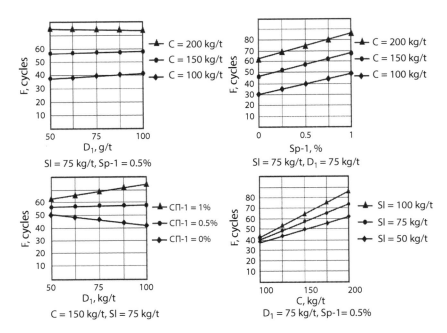

FIGURE 3.17 The effect of the composition of the dry mixture on the frost resistance of mortars.

The frost resistance of mortar on composite binders, as follows from the analysis of the regression equations (Table 3.19), is in the range of values allowed for masonry mortars and is mainly determined by factors affecting the binder-water ratio (i.e., cement consumption and the content of the additive of superplasticizer; Figure 3.17). The effect of the dust additive on the frost resistance of the mortar depends on how much the water content and, accordingly, the water binder ratio change when it is introduced. If there is no superplasticizer additive in the mixture, the introduction of a high dust content leads to a decrease in frost resistance; in the presence of the SP-1 additive, the frost resistance possibly cannot change or might even increase slightly.

The effect of alkaline oxides and free lime in the dust of the clinker kilns on the strength of mortars: In connection with possible fluctuations in the concentration of alkaline oxides and free lime in kiln clinker dust, special studies have been carried out on their effect on the properties of mortars based on composite binders, including Portland cement, blast furnace slag, and dust. For comparison, the physico-mechanical properties of solutions based on a binder containing only Portland cement and dust were also determined. A binder was obtained by mixing all the components in a laboratory paddle mixer. The slag was pre-crushed to a specific surface $S_{sp} = 310$ m²/kg to 330 m²/kg in a laboratory ball mill. Superplasticizer SP-1 in an amount of 1% by weight was introduced into the composition of the binders. The regulation of alkali and free lime in the binder was achieved by the addition of potassium carbonate and quicklime. The content of dust and slag filler in the binder

TABLE 3.22
Experiment Planning Conditions

Factors		Varying Settings	
Natural	Coded	−1	+1
R_2O^a (%)	x_1	2.5	5.5
CaO_{CB}^b (%)	x_2	2.8	5.8
Dust (%)^c	x_3	20	50
Dust/slag^d		0.3	1.2

^{a} The content of alkaline oxides in the binder.
^{b} The content of free calcium oxide in the binder.
^{c} Dust content (D_1) in the composition of binder No. 2.
^{d} The ratio of dust (D_1) to slag in binder No. 1.

was 50% by weight (binder No. 1). In a binder containing only dust as a mineral filler (binder No. 2), the amount of the latter ranged from 20% to 50%. Dry mixtures were obtained by adding three parts by weight of sand with fineness modulus $M_f = 1.3$ to the binders and additional mixing.

The experiments were carried out in accordance with the plan of a full-factor experiment 2^3. The experimental planning conditions are given in Table 3.22. Mortar mixtures were obtained from dry mixtures by adding water to a mobility of 8 cm to 10 cm by immersion of the cone.

As a result of processing and statistical analysis of the experimental data (Table 3.23), mathematical models (Table 3.24) of compressive and bending strength of hardened mortars were obtained based on the studied composite binders in the form of polynomial regression equations.

The influence of individual factors on the strength of mortars based on the studied composite binders resulting from the obtained models is shown in Figures 3.18 and 3.19. At the same time, on each of the graphs, the influence of two factors is considered, the third factor according to Table 3.22 is at the zero level.

Analysis of the strength dependencies shows that the degree and nature of the influence of the content of alkaline oxides and free lime in binders No. 1 and No. 2 is significantly different. An increase in the alkali content affects both the CaO content and the amount of clinker dust in binders. For mortars that do not contain blast furnace slag, an increase in the content of alkaline oxides in the composition of binders reduces strength. The increased alkali content is most negatively indicated for mortars with a relatively low content of free calcium oxide. The strength of the mortars decreases quite sharply with an increase in the composition of the binder dust content.

For binders containing a composite dust and slag filler, its effect on strength is determined by the composition of the dust and the mass ratio of dust:slag.

For mortar on composite binder No. 2, an increase in the alkali content from 3% to 7% (Figure 3.19) with a CaO_f content of 3% in 7 days causes a decrease in

TABLE 3.23
Strength Indicators of Mortars Based on Composite Binders

No.	x_1	x_2	x_3	W/B	Bending Strength (MPa in Days) 7	Bending Strength (MPa in Days) 28	Compressive Strength (MPa in Days) 7	Compressive Strength (MPa in Days) 28
1	+1	+1	+1	0.45	2.4	4	5.9	8.6
				0.41	4.8	6.9	12.1	17.8
2	+1	−1	+1	0.43	4.3	6.2	11.5	16.2
				0.39	5.2	6.8	11.3	17.2
3	−1	+1	+1	0.42	5.9	6.5	15.6	17.2
				0.39	5.4	8	12.4	20
4	−1	−1	+1	0.41	7.1	7.6	20.5	28.4
				0.38	5.9	8.4	15.4	20
5	+1	+1	−1	0.44	4.4	5.1	8.1	12.1
				0.4	4.5	8	8.9	17
6	+1	−1	−1	0.42	6.7	7.4	18.6	22.6
				0.39	6.4	9	13.8	19.5
7	−1	+1	+1	0.43	6.7	7.1	18.3	22
				0.38	6.4	8.7	13.7	19.3
8	−1	−1	−1	0.43	6.1	6.6	17	18.7
				0.39	5.9	8.8	22.3	30.4

Note: Above the line: water-binder ratio (W/B) and strength indicators for mortars based on composite binder No. 2; below the line: for binder No. 1.

compressive strength by 10%, and in 28 days by 20%. A similar negative effect on the strength has a change in the accepted range of variation and the content of free calcium oxide. At 7 days, with a simultaneous increase in the content in binder No. 2 of alkali and CaO_{free}, their negative effect is summarized. For samples of 28 days, the negative effect of alkalis in the binder with an increased content of free lime is reduced.

The most significant negative effect on the strength of mortars has an increase in the composition of the binder dust content. With an increase in the content of clinker dust with an increased amount of alkali up to 50% by weight of the binder, the compressive strength of the mortars decreases by 2.5 times.

The negative effect of alkalis when clinker dust is introduced into the binder is significantly weakened if blast furnace slag is present (Figure 3.18). With an increase in the content of free lime in the composite slag-containing binder, the strength of the mortars remained practically unchanged (Figure 3.18), and even increased with a relatively low alkali content. With an increase in the dust:slag ratio, a slight decrease

TABLE 3.24
Experimental-Statistical Models of Strength Properties of Mortars Based on Composite Binders

Strength Mortar (Days)			Statistical Models	
Bending Strength (MPa)		7 days	$y'_1 = 5.56 - 0.0375 \cdot x_1 + 0.0125 \cdot x_2 - 0.5875 \cdot x_3$ $+ 0.063 \cdot x_1 x_2 - 0.288 x_1 x_3 + 0.013 x_2 x_3$	(3.31)
			$y_1 = 5.45 - 0.4 \cdot x_1 \quad 0.575 \cdot x_2 - 1.2 \cdot x_3$ $+ 0.075 \cdot x_1 x_2 - 0.450 x_1 x_3 - 0.325 x_2 x_3$	(3.32)
		28 days	$y'_2 = 8.08 + 0.075 \cdot x_1 - 0.2 \cdot x_2 - 0.65 \cdot x_3$ $- 0.05 x_1 x_3 - 0.375 x_2 x_3$	(3.33)
			$y_2 = 6.31 - 0.4125 \cdot x_1 - 0.2625 \cdot x_2 - 0.8625 \cdot x_3$ $+ 0.063 \cdot x_1 x_2 - 0.488 x_1 x_3 - 0.088 x_2 x_3$	(3.34)
Compressive Strength (MPa)		7 days	$y'_3 = 13.74 - 1.6125 \cdot x_1 + 1.1375 \cdot x_2 - 2.5625 \cdot x_3$ $- 0.313 \cdot x_1 x_2 + 0.938 x_1 x_3 - 0.613 x_2 x_3$	(3.35)
			$y_3 = 14.44 - 1.7125 \cdot x_1 - 1.1875 \cdot x_2 - 4.1625 \cdot x_3$ $+ 0.713 \cdot x_1 x_2 - 1.563 x_1 x_3 - 0.388 x_2 x_3$	(3.36)
		28 days	$y'_4 = 20.15 - 1.75 \cdot x_1 + 1.075 \cdot x_2 - 2.15 \cdot x_3$ $- 0.825 \cdot x_1 x_2 + 1.15 x_1 x_3 - 1.575 x_2 x_3$	(3.37)
			$y_4 = 18.23 - 1.9 \cdot x_1 - 1.7 \cdot x_2 - 4.7 \cdot x_3 + 0.975 \cdot x_1 x_2$ $- 1.275 x_1 x_3 + 0.575 x_2 x_3$	(3.38)

Note: $y_1 - y_4$ are the models of strength properties of mortar based on composite binder No. 2; $y'_1 - y'_4$ binder No. 1.

in the strength of mortar is noted, which weakens with an increased alkali content. At the maximum dust and slag ratio with an increase in the content in the binder alkalis and free lime, a decrease in strength is practically not observed. Thus, the performed studies show that in mortars on a composite cement-dust slag binder, a significant content of alkalis and free lime can have a positive effect on the strength of the mortar, which is obviously due to their activating effect on the slag component of the binder. In this case, the choice of the optimal mass ratio of slag and dust, considering the chemical composition of the latter, is essential.

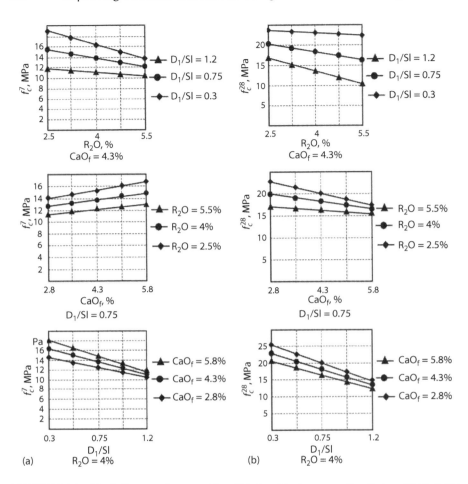

FIGURE 3.18 The influence of compositional binders No. 1 composition factors (Table 3.23) on the strength of mortars: (a) 7 days and (b) 28 days (R_2O, CaO_f, D_1, Sl—content of alkali, free calcium oxide, clinker dust, blast furnace slag, respectively).

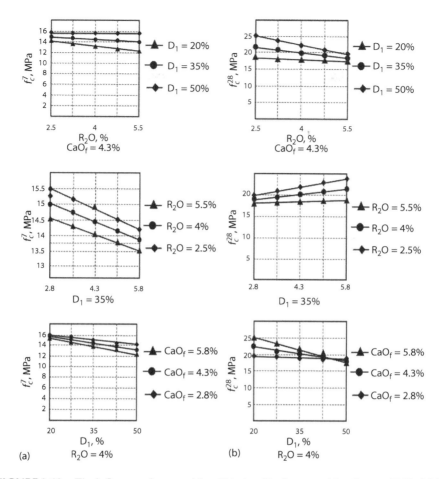

FIGURE 3.19 The influence of compositional binders No. 2 composition factors (Table 3.23) on the strength of mortars: (a) 7 days and (b) 28 days (R_2O, CaO_f, D_1, Sl is the content of alkali free calcium oxide, clinker dust, blast furnace slag. respectively).

References

1. Thomas, M. 2014. Optimizing the use of fly ash in concrete. *Portland Cement Association*, 2, 1–24.
2. Galyna Kotsay. 2016. Pozzolanic activity diagnostics of fly ash for Portland cement. *Chemistry & Chemical Technology*, 10(3), 355–360.
3. Berry, E.E., Malhotra, V.M. 1982. Fly ash for concrete. Critical Review. *ACI Journal*, 2(3), 59–73.
4. Dvorkin, L.I., Solomatov, V.I., Vurovoy, V.N., Chudnovskiy, S.M. 1991. *Cement Concrete with Mineral Fillers*. Budivelnyk, Kiev, 137 p. (in Ukrainian).
5. Sanitsky, M.A., Soboľ, C.H.S., Markiv, T.E. 2010. *Modified Composite Cements*. Lvov, Polytekhnika. (in Ukrainian).
6. Thomas, J.J., Jennings, H.M., Chen, J.J. 2009. Influence of nucleation seeding on the hydration mechanisms of tricalcium silicate and cement. *Journal of Physical Chemistry C*, 113(11), 4327–4334. doi:10.1021/jp809811w.
7. Krivenko, P.V. 1992. *Special Slag-Alkali Cements*. Kyiv: Budivelnik, 192 p. (in Ukrainian).
8. Rai, B., Kumar, S., Satish, K. 2014. Effect of fly ash on mortar mixes with quarry dust as fine aggregate. *Advances in Materials Science and Engineering*, Article ID 626425, 7 p.
9. Thandavamoorthy, T.S. Feasibility of making concrete using lignite coal bottom ash as fine aggregate. *Archives of Civil Engineering*.
10. Bacikova, M., Stevulova, N. 2009. Possibilities of coal fly ash utilization in the highway engineering, Chania, Crete, Greece, CEST, b-88–b-95p.
11. Bacikova, M., Stevulova, N. Examination of fly ash utilization suitability for the production of cement–concrete cover of pavement. *Chemine Technologija*, 1(50). Kaunas, Lithuania, 24–29.
12. Dinakar, P., Babu, K.G., Santhanam, M. 2008. Durability properties of high volume fly ash self compacting concretes. *Cement & Concrete Composites*, 30, 880–886. doi:10.1016/j.cemco-dcomp.2008.06.011.
13. Ondova, M., Stevulova, N., Fecko, L. 2010. Fly ash influence on the chemical and mechanical properties of cement concrete cover of pavement. Yucomat, Montenegro, 164 p.
14. Hewlett, P.C. 2004. *Lea's Chemistry of Cement and Concrete*, 4th edition. Butterworth-Heinemann, Oxford, 1092.
15. Sychev, M.M. 1987. Prospects for increasing the strength of cement stone. *Cement*, (9), 17–19. (in Russian).
16. Dvorkin, L.I., Dvorkin, O.L., Korneychuk, Y.A. 1998. *Effective Cement-Ash Concrete*. Rovno, 195 p. (in Ukrainian).
17. Xu, Z., Linzhy, L. 1986. Research on superfine fly ash and its activity. *Beijing International Symposium on Cement and Concrete*, Beijing, China, May 14–17, 1, 493–507.
18. Lane, R.O, Best, S.F. 1982. Properties and use of fly ash in Portland cement concrete. *Concrete International*, 4(7), 81–92.
19. Kurdowski, W. 2010. Chemia Cementu i Betonu. Stowarzyszenie producentów cementu. PWN, Warszawa.
20. Bazhenov, Y.M., Demyanova, V.S., Kalashnikov, V.I. 2006. *Modified High-Quality Concrete*. Moscow, 368 p. (in Russian).

21. Stolnikov, V.V. 1989. Use of fly ash from the combustion of pulverized fuel in thermal power plants. 50 p. (in Russian).
22. Ramachandran, V.S., Feldman, R.F., Kollepardi, M. et al. 1988. *Additives in Concrete: Reference Book*. Moscow, Stroyizdat, 575 p. (in Russian).
23. Kishore, K. 1997. Concrete mix design with fly ash & superplasticizer. *ICI Bulletin*, 29–30.
24. Eriksen, K., Nepper-Christensensen, P. 1981. Experiences in the use of superplasticizers in some special fly-ash concretes. *Special Publication*, 68, 1–20.
25. Rebinder, P.A. 1961. *Surfactants*. Moscow, Knowledge, 116 p.
26. Ramachandran, V.S. 1983. Adsorption and hydration behavior of tricalcium aluminate–water and tricalcium aluminate–gypsum-water systems in the presence of superplasticizers. *Journal of the American Concrete Institute*, 80, 235–241.
27. Ratinov, B.B., Rosenberg, T.I. 1989. *Admixtures in Concrete*. Moscow, Stroyizdat, 188 p. (in Russian).
28. Roy, D., Daimon, M. 1980. Effect of Admixtures upon electrokinetic phenomena during hydration of C3S.C3A and Portland cement. *7th International Congress on the Chemistry of Cements*, Paris, France, vol II, 242–246.
29. Lowke, D., Gehlen, C. 2017. The zeta potential of cement and additions in cementitious suspensions with high solid fraction. *Cement and Concrete Research*, 95, 195–204.
30. Dvorkin, L.I., Dvorkin, O.L. 2015. *Effective Cement-Ash Concrete and Mortar*. Monograph. Palmarium Academic Publishing, 436.
31. Dvorkin, L., Dvorkin, O. 2006. *Basics of Concrete Science*. Stroybeton, St. Petersburg, 686 p. (in Russian).
32. Gambhir, M.L. 2013. *Concrete Technology: Theory and Practice*. Tata McGraw-Hill Education, New Delhi, India. 792 p.
33. Batrakov, V.G. 1998. *Modified Concrete. Theory and Practice*. Moscow, 768 p. (in Russian).
34. Marushchak, U., Sanytsky, M., Mazurak, T., Olevych, Y. 2016. Research of nanomodified Portland cement compositions with high early age strength. *Eastern-European Journal of Enterprise Technologies*, 6(6), 50–57.
35. Scrivener, K.L., Nonat, A. 2011. Hydration of cementitious materials, present and future. *Cement and Concrete Research*, 41(7), 651–665. doi:10.1016/j.cemconres.2011.03.026.
36. Wang, Q., Yan, P.Y., Feng, J.J. 2013. Design of high-volume fly ash concrete for a massive foundation slab. *Magazine of Concrete Research*, 65(2), 71–81.
37. Papadakis, V.G. 1999. Effect of fly ash on Portland cement systems. Part I. Low-calcium fly ash. *Cement and Concrete Research*, 29(11), 1727–1736.
38. Butt, Y.V., Timashev, V.V. 1973. *Workshop on the Technology of Binders*. Moscow Higher, Shk, 504 p. (in Russian).
39. Dvorkin, L., Dvorkin, O., Ribakov, Y. 2012. *Mathematical Experiments Planning in Concrete Technology*. Nova Science Publishers, New York, 173 p.
40. Kolmykova, E.E., Mikhailov, N.V. 1957. Study of the processes of structure formation in cement paste. *Concrete and Reinforced Concrete*, (4), 118–126.
41. Dvorkin, L.Y., Dvorkin, O.L., Yu, Garnitsky, V., Rizhenko, I.M. 2013. *Modified Ash-Containing Dry Mortars for Masonry and Adhesive Mortars*. Monograph. Rivne, 219 p. (in Ukraine).
42. Ahverdov, I.N. 1981. *Basis of Concrete Physics*. Stroyizdat, Moscow. 464 p. (in Russian).
43. Justs, J., Bajare, D., Korjakins, A., Mezinskis, G., Locs, J., Bumanis, G. 2013. Microstructural investigations of ultra-high performance concrete obtained by pressure application within the first 24 hours of hardening. *Construction Science*, 14, 50–57.
44. Sanchez, F., Sobolev, K. 2010. Nanotechnology in concrete—A review. *Construction and Building Materials*, 24, 2060–2071.

45. Biricik, H., Sarier, N. 2014. Comparative study of the characteristics of nano silica-, silica fume and fly ash-incorporated cement mortars. *Materials Research*, 17, 570–582.
46. Goldman, A., Bentur, A. 1993. The influence of microfillers on enhancement of concrete strength. *Cement and Concrete Research*, 23, 962–972.
47. Robert, R., Schneider, W., Dickey, L. 1987. *Reinforced Masonry Design*. Prentice Hall, Upper Saddle River, NJ, 682 p.
48. Zhao, H., Deng, M. 2011. Effects of the synthetic ways of sulfonated-phenol-formaldehyde (SPF) polymers on application properties of cementitious materials. *Polymer-Plastics Technology and Engineering*, 50(9), 916–922. doi:10.1080/036025 59.2011.551983.
49. Szwabowski, J., Łaźniewska-Piekarczyk, B. 2008. The increase of air content in SCC mixes under the influence of carboxylate superplasticizer. *Cement, Wapno, Beton*, 4, 205.
50. Gebler, S., Klieger, P. 1983. Effect of fly ash on the air-void stability of concrete. Fly ash, silica fume, slag and other mineral by-products in concrete. *Publication ACI*, 795, 103–142.
51. Bazhenov, Y.M. 1987. *Concrete Technology*. Vysshaya shkola, Moscow, 449 p. (in Russian).
52. Basin, V. 1981. *Adhesive Strength*. Chemistry, Moscow, 208 p. (in Russian).
53. Aitcin, P.-C. 1998. *High-Performance Concrete*. Modern Concrete Technology, E & FN Spon, London, UK, 591 pp.
54. Kalashnikov, V., Demyanova, N. 2001. *Polymer Mineral Dry Building Mixtures*. News of Higher Educational Institutions, Building, 5, 41–46. (in Russian).
55. Dvorkin, L., Dvorkin, O. 2006. *Basics of Concrete Science: Optimum Design of Concrete Mixtures*. Amazon. (Kindle edition)/(e-book), Stroi-Beton, S-Peterburg, 382 p. (in Russian).
56. Kosmatka, S.H., Wilson, M.L. 2011. *Design and Control of Concrete Mixtures*, 15th edition. Portland Cement Association, Skokie, 460 p.
57. Batrakov, V.G. 1990. *Modified Concretes*. Stroyizdat, Moscow, 396 p. (in Russian).
58. Stolnikov, V.V., Litvinov, R.E. 1972. *Fracture Resistance of Concrete*. Moskow, Energy, 113 p. (in Russian).
59. Neville, A.M. 1996. *Properties of Concrete*, 4th edition. John Wiley & Sons, New York, 844 p.
60. Gorchakov, G.I., Kapkin, M.M., Skramtaev, B.G. 1965. *Increasing the Frost Resistance of Concrete in the Structures of Industrial and Hydraulic Structures*. Moskow, Stroizdat, 196 p. (in Russian).
61. Dvorkin, L., Gotz, V., Dvorkin, O. 2014. *Testing of Concrete and Mortar*. Design composition, 397 p. (in Ukrainian).
62. Mohajerani, A., Suter, D., Jeffrey-Bailey, T., Song, T., Arulrajah, A., Horpibulsuk, S., Law, D. 2019. Recycling waste materials in geopolymer concrete. *Clean Technologies and Environmental Policy*, 21(3), 493–515.
63. Chatziaras, N., Psomopoulos, C., Themelis, N. (2016). Use of waste derived fuels in cement industry: A review. *Management of Environmental Quality*, 27(2), Guide for Specifying Mixing 178–193. doi:10.1108/MEQ-01-2015-0012.
64. Swamy, R. 1997. Design for durability and strength through the use of fly ash and slag in concrete. *Proceedings of the Mario Collepardi Symposium on Advances in Concrete Science and Technology*, 127–194.
65. Singh, M. 1993. Use of industrial by-products in the manufacture of cements and alternative binders. *Indian Cement Review-Annual' 92'*, 119–128.
66. Dvorkin, L., Dvorkin, O., Ribakov, Y. 2015. *Construction Materials Based on Industrial Waste Products*. New York, Nova Publishers. 252 p.

67. Barnes, H.L. (Ed.) (1979) *Geochemistry of Hydrothermal Ore Deposits*, 2nd Edition. Wiley-Interscience, New York, 798.
68. Richardson, I.G., Wilding, C.R., Dickson, M.G. 1989. The hydration of blast furnace slag cements. *Advances in Cement Research*, 2(8), 147–157.
69. Mirkin, L. 1961. *Handbook of X-ray Structural Analysis of Polycrystals*. Moscow: State Publishing House of Physical and Mathematical Literature, 863 p. (in Russian).
70. Runova, R.F., Binders R.F., Runova, L.J. Dvorkin, O.L., Dvorkin, Y.L. 2012. Nosovsky. - Kyiv: Basis, 448 p. (in Ukrainian).
71. Abbas, M.A., Jarullah, D.S. 2016. Producing lightweight foam concrete building units using local resources. *Civil and Environmental Research*, 8(10), 54–62.
72. Amran, Y.M., Farzadnia, N., and Ali, A.A. (2015). Properties and applications of foamed concrete: A review. *Construction and Building Materials*, 101(1), 990–1005. doi:10.1016/j.conbuildmat.2015.10.112.
73. Dvorkin, L. 2013. Technology and properties of foam concrete based on fine-ground binder. L. Dvorkin, O. Bordyuzhenko Building materials, products and sanitary engineering. *Scientific and Technical Collection*. Issue 49, K.: NDIBMV 2013, 49–53.
74. Hudson, K. (1991). Foamed concrete for trench reinstatement. *New Zealand Concrete Construction*, 35(1), 6–9.
75. Videla, C., Lopez, M. (2000). Mixture proportioning methodology for structural sand-lightweight concrete. *ACI Materials Journal*, 97(3), 281–289.
76. Aldridge, D. 2005. Introduction to foamed concrete: What, why, how? In: Dhir, R.K., Newlands, M.D., McCarthy, A., editors. Use of Foamed Concrete in Construction. London, UK, Thomas Telford, 1–14.
77. Puttappa, C.G. 2008. Mechanical properties of foamed concrete. *International Conference on Construction and Building Technology*, 43, 491–500.
78. Lim, B.Y. 2006. Study of water ingress into foamed concrete, National University of Singapore, Master's Thesis, Singapore.
79. Karl, S., Worner, J.-D. 1994. Foamed concrete–mixing and workability. *Proceedings of the International RILEM workshop Special Concretes; Workability and Mixing* (Ed. P J M Bartos), held at Paisley, Scotland, 217–224.
80. Andreev, A. 2014. Influence of particle size distribution of crushed wood for wood-cement material on its strength. *Engineering Science from Theory to Practice* (32), 71–76. (in Russian).
81. Chowdhury, S., Maniar, A., Suganya, O.M. 2015 Strength development in concrete with wood ash blended cement and use of soft computing models to predict strength parameters. *Journal of Advanced Research*, 6, 907–913.
82. Hermawan, D., Subiyanto, B., Kawai, S. 2001b Manufacture and properties of oil palm frond cement-bonded board. *Journal of Wood Science*, 47, 208–213.
83. Lee, A.W.C., Hong, Z. 1986. Compressive strength of cylindrical samples as an indicator of wood-cement compatibility. *Forest Products Journal*, 36(11/12), 87–90.
84. Bazhenov, Y.M. 1888. Increasing the efficiency and economy of concrete technology. *Concrete and Reinforced Concrete* (9), 14–16.
85. Guide for Specifying Mixing, Placing and Finishing Steel Fiber Reinforced Concrete. *ACI Committee*, 544, 1993, ACI Mater.
86. Rabinovich, F.N. 1995. *Concretes with Dispersed Reinforcement*. Taylor & Francis Group, New York, 214 p.
87. Fuat, K., Faith, A., Iihami, Y., Yusa, S. 2008. Combined effect of silica fume and steel fiber on the mechanical properties of high strength concretes. *Construction and Building Materials*, 22, 1874–1880.
88. Bayramov, F., TasDemir, C., TasDemir, M.A. 2004. Optimisation of steel fiber reinforced concretes by means of statistical response surface method. *Cement and Concrete Research*, 26, 665–675, 2004.

References

89. Job, T., Ananth, R. 2007. Mechanical properties of steel fiber-reinforced concrete. *Journal of Materials in Civil Engineering*, 19(5), 385–392.
90. Gorchakov, G.I. 1965. *Increasing the Frost Resistance of Concrete in the Structures of Industrial and Hydraulic Structures*. G.I. Gorchakov, M.M. Kapkin, B.G. Skramtaev. Moskow: Stroizdat 195 p.
91. Cohen, M.D., Zhou, Y., Dolch, W.L. 1992. Non-air-entrained high strength concrete-is it frost resistant. *ACI Materials Journal*, 89(4), 406–415.
92. Li, Y., Langan, B.W., Ward, M.A. 1994. Freezing and thawing: Comparison between non-air-entrained and air-entrained high-strength concrete. *ACI Special Publication*, 149.
93. Runova, R.F. *The Technology of Modified Mortars*. R.F. Runova, Y. L. Nosovsky. - K: KNUBiA Publishing House, 2007. 256 p. (in Ukrainian).
94. Zakharchenko, P.V. *Modern Composite Building and Finishing Materials*. Zakharchenko, P.V., Dolgiy, E.M., Galagan, Y.O. KNUBiA Publishing House, 2005. 512 p. (in Ukrainian).
95. Bouras, R., Mohand, CSIH, Sonebi, M. Adhesion and rheology of joints fresh mortars. *Journal of Materials and Engineering Structures*, 6(2), 157–165.
96. Zhu, Y., Ma, B., Li, X., Hu, D. 2013. Ultra high early strength self-compacting mortar based on sulfoaluminate cement and silica fume. *Journal of Wuhan University of Technology-Materials Science*, 28(5), 973–979.
97. Butt, Y.M., Sychev, M.M., Timashev, V.V. 1980. *Chemical Technology of Binder Materials*. Moskow: Higher School. 472 p. (in Russian).
98. Beltagui, H., Sonebi, M., Maguire, K., Taylor, S. 2018. Feasibility of backfilling mines using cement kiln dust, fly ash, and cement blends. *MATEC Web of Conferences*, 149, 01072.
99. Forinton, J. 2013. Recycling kiln bypass dust into valuable materials. *IEEE-IAS/PCA, Cement Industry Technical Conference*, Orlando, 240 p.
100. Dvorkin, L., Dvorkin, O. 2007. *Building Materials from Industrial Waste*. Rostov-on-Don, Phoenix. 369 p.
101. Smaouia, N. 2005. Effects of alkali addition on the mechanical properties and durability of concrete. *Cement and Concrete Research*, 35, 203–212.
102. Gidley, S.J., Sack, A.W. 1984. Environmental aspects of waste utilization in construction. *ASCE-Journal of Environmental Engineering*, 110.
103. Menendez, G., Bonavett, V., Irassar, E.F. (2003). Strength development of ternary blended cement with limestone filler and blast-furnace slag. *Cement &Concrete Composites*, 25, 61–67.
104. Volzhenskiy, B., Sudakas, L.H. 1980. *Handbook of Cement Chemistry*, Moskow, 224. (in Russian).
105. Dvorkin, L.I. 2011. *Building Mineral Binders*. L.I. Dorkin, and O.L. Dorkin. Moskow: Infra-Engineering. 544 p.
106. Henning, O., Shtiler, R. Use of high-alkaline cement dust in Portland cement. *Proceedings of the 6th International Congress on the Cement Chemistry*, Moskow, 1974, 27. (in Russian).
107. Boldyryev, A.S. 1989. Use of waste and secondary resources in the building materials industry. *Building Materials*, 7(5). (in Russian).
108. Gidley, S.J., Sack, A.W. 1984. Environmental aspects of waste utilization in construction. *ASCE-Journal of Environmental Engineering*, 110.
109. Kuatbaev, K.K. 1981. *Silicate Concretes from Industrial By-Products*, Moskow, 246. (in Russian).
110. Suleymenov, A.T. 1986. *Binding Materials from Industrial By-Products*, Moskow, 189 p. (in Russian).
111. Ternovaya, E.A. 1987. Use of dust from cement plants rotary kilns. *Constructing Materials*, 10, 22. (in Russian).

112. Beltagui, H., Sonebi, M., Maguire, K., Taylor, S. 2017. Utilisation of cement kiln dust for the activation of fly ash in low strength applications. *2-nd ICBBM*, 5 p.
113. Alekhin, Y.A., Lyusov, A.M. 1988. *Economic efficiency of the use of secondary resources in the production of building materials.* Moskow: Stroyizdat, 344 p. (in Russian)
114. Buchwald, A., Schulz, M. 2005. Alkali-activated binders by use of industrial by-products. *Cement and Concrete Research*, 35, 968–973.
115. Kirschner, A.V., Harmuth, H. 2004. Investigation of Geopolymer binders with respect to their application for building materials. *Ceramics – Silikaty*, 3, 117–120.
116. Glukhovskyy, V., Makedon, N. Alkaline cements containing clinker kilns dust, Alkaline cements, concretes and structures, 1989, 35. (in Russian).
117. Aitcin, P.C. 2014. *Binders for Durable and Sustainable Concrete.* CRC Press, Boca Raton, FL, 528 p.
118. Peethamparan, S., Olek, J., Lovell, J. 2008. Influence of chemical and physical characteristics of cement kiln dusts (CKDs) on their hydration behavior and potential suitability for soil stabilization. *Cement and Concrete Research*, 38, 803–815.
119. Guidelines for the design of road garments with bases made of materials reinforced with inorganic binders. Moskow: SoyuzdorNII, 1985 – 25 p. (in Russian).
120. Guidelines for the use of low quality stone materials and contaminated sands treated with inorganic binders. Moskow: SoyuzdorNII, 1990–31s. (in Russian).
121. Bazhenov, Y.M., Gorchakov, G.I., Alimov, L.A., Voronin, V.V. 1978. *Obtaining Concrete with Given Properties.* Stroyizdat, Moscow, 54 p. (in Russian).
122. Uriev, N., Dubinin, I. 1980. *Colloid-Cement Mortars.* Stroyizdat, St. Petersburg, 192 p. (in Russian).
123. Chana, P. 2011. Low carbon cements: The challenges and opportunities. In *Proceedings of the Future Cement Conference & Exhibition*, London, UK, February 8–9, 1–7.
124. Memon, A.H., Radin, S.S., Zain, M.F.M., Trottier, J.F. 2002. Effect of mineral and chemical admixtures on high-strength concrete in seawater. *Cement and Concrete Research*, 32, 373–377.
125. Powers, T.C., Brownyard, T.L. 1947. Studies and physical properties of hardened Portland cement paste. *Proceedings of the American Concrete Institute* 43(9), 249–336.
126. Dvorkin, L.I. 1981. *Optimal Design of Concrete Compositions.* Lviv: Higher School, p. 160. (in Russian).
127. Taylor, H.F. 1990. *Cement Chemistry.* London, UK: Academic Press, 360 p.
128. Kurdowski, W. 1991. *Chemia Cementu.* Warszawa: Wydawnictwo Naukowe PWN, 200 p.
129. Orhard, D.F. 1979. *Concrete Technology.* London, UK. Vols. 1, 2, 1033 p.
130. Burov, Y.S., Kolokolnikov, V.S. 1974. Laboratory Workshop on the course "Mineral Binders." Moskow, 172. (in Russian).
131. Pierce, C.E., Tripathi, H., Brown, T.W. 2003. Cement kiln dust in controlled low-strength materials. *ACI Materials Journal*, 100(6), 455.
132. He, J., Ca, J., Sun, Z. 2018. Hydration characteristics of alkali-activated slag cement. *Romanian Journal of Materials*, 48(8), 168.
133. Batrakov, V.G., Babaev, S.G., Bashlykov, N.F. 1998. Concretes on binders with low water demand. *Concrete and Reinforced Concrete*, (11), 4–6.
134. Collepardi, M., Corradi, M., Valente, M. 1981. Influence of polymerisation of sulfonated naphtaline condensate of the interaction with cement—Developments in the use of superplasticizers. *ACI*, 485–498.
135. Costa, U., Massazza, F., Barrila, A. 1982. Adsorption of superplasticizers on C3S, changes in zero potential and rheology of pastes. *Cemento*, 79(4), 323–336.

136. Massazza, F., Costa, U., Barrila, A. 1981. Adsorption of superplasticizers in calcium aluminate monosulfate hydrate. *Special Publication*, 68, 499–514.
137. Soroker, V.I., Dovzhik, V.G. 1964. *Stiff Concrete Mixtures in Precast Reinforced Concrete Production*. Moscow, Stroyizdat, Moscow, 206 p. (in Russian).
138. Wallevik, J.E. 2009. Rheological properties of cement paste: thixotropic behaviour and structural breakdown. *Cement and Concrete Research*, 39, 14–29.

Index

A

absorption, 2, 4, 7–9, 19–20, 70
acid-base reaction, 74
activating, 5–7, 22, 61, 64, 70, 72, 130, 146, 169
activator, 7, 62–64, 67, 69–70, 72, 74–75, 77–79, 114
active filler, 1
activity, 2, 5–7, 22, 33–34, 37, 62, 66, 70, 74, 76, 114, 129–130, 143, 148, 164
adhesive, 3, 6, 37–40, 104, 164–165
aggregate, 2–6, 26, 29, 41, 45, 52–53, 58–59, 82, 91, 96, 98, 102, 104, 108, 114, 121, 124, 129
aging, 53, 84
air entraining, 30, 109
algorithmic, 134
alite, 3, 18
aluminosilicate, 1–2
amorphization, 5
autoclave, 92, 95
average pore size, 21

B

ball bearings, 2
belite, 18
binder, 1–6, 11, 14, 19, 22–29, 32–33, 37–38, 40–42, 47, 51, 54, 56–57, 61–64, 66–67, 70, 72–76, 78–82, 85, 91–92, 96, 98–105, 107–109, 113–114, 117–120, 129–161, 163–171
blast furnace slag, 7, 9, 22, 41, 62–64, 74, 85, 95, 129, 132–133, 136, 148, 150, 154, 156, 162, 166–168, 170–171
bulk density, 20, 89, 103

C

calcite, 130
calcium chloride, 70, 96, 98
  hydrosilicate, 8, 19, 64, 74, 79, 81, 94
  hydrosulfoaluminate (ettringite), 18–19, 64, 73, 80–81
capillary, 3–4, 20–21, 26, 56–57
cast mixtures, 57
cement, 1–5, 7, 9–26, 28–33, 37–45, 47–48, 52–59, 61, 63–64, 66, 70, 72–79, 81, 85, 89, 91–94, 96–98, 100, 102, 104–106, 108–110, 113–115, 117–118, 121–122, 124, 127, 129–136, 140–146, 148–154, 156–166, 169
  paste, 5, 7, 13, 15, 17, 25–26, 40, 53, 55–56, 72–73, 77–79, 117, 165
  stone, 1–2, 18–21, 37, 52–53, 55–56, 79, 81, 93–94, 96, 98, 113–114, 143, 145–146
cementing efficiency, 3
chemical method for activating, 5, 9
  activity, 37, 143
chemisorptions, 6, 38
clay, 2, 127–129
clinker, 1, 3, 17, 19, 22, 41, 47, 61, 63–64, 66–68, 70, 72–74, 76, 82, 91–92, 95–96, 98, 106–107, 109, 114, 120, 127–133, 136, 142, 146, 150, 153–154, 164–168, 170–171
  kiln dust, 127, 136, 146
coagulation, 1–2, 14, 16
  contacts, 15
cold bridges, 121
colloidal, 5, 8
composition design, 3, 57–58, 101, 118, 161
compressive strength, 4, 22–23, 33, 44, 46–50, 53–54, 56, 58, 62, 65, 68–73, 78, 82–84, 87, 89, 91–93, 95–97, 99, 105, 108, 114–115, 118–124, 130, 147–152, 157–164, 168
concrete mixture stratification, 131
cone flow on shaking table, 27, 33, 72, 101
cone slump, 90
contraction, 4
corrosion, 4, 109–114, 129–130
  resistance, 106, 108, 110
crack, 6, 33, 44, 54–55, 96
  resistance, 33, 44, 54–55
cramped conditions, 2, 85
creep, 4, 54, 98
crushed stone, 58–59, 82, 89, 91, 98, 103–104, 108
crystalline structure, 5, 63
crystallization, 14–16, 63, 78, 142
cyclone, 128

D

defect, 56
deforming properties, 52
degree of filling, 6, 15
diffraction pattern, 18, 79–81
dispersion, 2–3, 6–7, 9, 11, 29, 32–33, 37, 42–44, 47–48, 56–58, 63, 67, 73–74, 79, 98, 128–129, 134, 139, 153, 164

dry construction mixtures (DCM), 114, 121, 127, 156
dust, 6, 127–138, 140–142, 144, 146–154, 156–170

E

economic effect, 7
efflorescences, 64, 67
elasticity, 4, 52–53, 96
electrical conductivity, 17, 62
elongation, 142
experimental-statistical models, 31, 33, 35, 39, 73, 83–84, 91–93, 96–97, 99, 101, 105, 108, 115, 117–118, 122, 124, 161, 169
expressing, 89

F

factor space, 137
fiber, 98–109
fibrous structures, 81
fine-grained concrete, 11, 100–101
fineness modulus, 82, 114, 167
firing mode, 127–129, 131
fluctuation, 131, 142, 153, 166
fly ash, 1–2, 4–5, 7–11, 13, 15, 17, 19, 22, 25–33, 36, 39–41, 54, 56–58, 61, 156–157
foam concrete, 92–95
forming, 5, 30, 64, 121
fraction, 2–3, 24, 128, 131
freezing and thawing, 106
frequencies, 142
frost resistance, 4–5, 44, 56–57, 106, 108–109, 130, 157–159, 161, 166
furnace, 127–129

G

gel-crystallite phase, 52
granite, 41, 58, 82, 98, 108
grinding, 2–3, 5–7, 13, 17, 22, 24–25, 30–31, 37, 40–41, 61, 64–67, 92, 131–135, 137, 139, 142, 144, 146, 148, 153–156

H

hardening accelerators, 71
hardening conditions, 57
heat treatment, 49, 51–52, 57, 63, 82, 84–85, 87, 89–95
high performance concrete, 40
high-strength, 98
hydrated, 1–2, 9, 15, 18, 37, 53, 63, 79–80, 116–117, 120, 129, 133, 143–144
hydration degree, 10

hydraulic activity, 2, 129–130
hydrophilic, 164
hydrosilicate, 8, 16, 19, 62–64, 70, 74, 79, 81, 94, 130
hydroxyl ions, 30

I

induction period, 14–15, 17, 79
involved air, 30
isothermal exposure, 50–51, 84

K

kiln, 127–132, 147, 154, 165–166

L

light aggregate, 121
lignosulfonate, 86
lime, 7–9, 63, 76, 78, 82, 107, 114–116, 118, 127, 129, 146, 153, 156–157, 166–167, 169
limestone, 2, 127, 165
liquid phase, 30
loss on ignition (LOI), 130–131
low-clinker cement, 61, 72, 92, 106, 114, 120
low clinker slag Portland cement (LSC), 64–99, 101–104, 106–124

M

mechanical method for activating, 5
mechanochemical method for activating, 5
metakaolin, 7
micro-concrete, 37
mineralogical composition, 2–3, 19, 81
modulus of elasticity, 4, 52–53
mortar, 1–5, 13, 22, 24, 26–37, 39–40, 66, 70, 72–73, 75–77, 95–96, 109–110, 114–119, 121–125, 129–131, 134, 146, 151–152, 156–171

N

naphthalene-formaldehyde superplasticizer, 95
nomogram, 43–44, 48–49, 57–58, 101–103, 118, 161–162
normal conditions, 9, 46, 62–63, 91, 142–143
normal consistency, 22–23, 25–27, 44–45, 65, 68, 72–73, 117, 131–135, 139, 143, 145–146
normal-weight concrete, 82, 106

O

output parameter, 157

Index

P

peptization, 5
phase composition of cement stone, 19
phosphogypsum (FG), 64, 66–67, 69–70, 72–74, 82
plasticizers, 5, 66, 91, 134
plastogram, 15, 77–78, 139, 142
polyacrylate, 92
polycarboxylate, 8, 10–11, 13, 22, 25–26, 28–31, 33, 40–41, 125, 134, 152
polyfunctional modifier (PFM), 7, 9–21, 25–44, 46, 48–50, 55, 57–59
polynomial, 13, 25, 33, 47, 72, 82, 92, 98, 114, 119, 123, 148, 167
pore size uniformity, 21
porosity, 4, 18–21, 52–53, 55–57, 67, 93, 108, 113, 121, 146
powder, 5–7, 24, 37, 128, 143, 146, 154, 157
powdered, 6, 157
pozzolanic activity, 1, 7–9
propylene glycol, 24, 64–66, 154

Q

quartz sand, 41, 114, 121, 157, 165

R

recrystallization, 64, 67, 79
regression equations, 13, 25, 33, 45, 47, 49, 72, 82, 92, 98, 104, 114, 123, 136, 138, 157, 159, 166, 167
regrinding, 150, 153
reinforced concrete structures, 49, 64, 76, 108
reinforcement, 98, 100, 106, 108–111, 113–114, 130

S

sand, 27–28, 33, 41, 58–59, 62, 66, 72, 82, 89, 91–93, 98, 101, 103–104, 108–110, 114, 121, 157–158, 162–163, 165, 167
sawdust concrete, 96–97
separation, 3, 8, 29–30, 55–56, 116–117, 129
shrinkage deformation, 4, 54
sieve, 143
silica fume, 7
single-stage grinding, 64, 66
slag alkali binders (SAB), 63–64, 130
sludge, 67, 128–129
sodium silicon fluoride, 70, 72–76, 79, 82, 107, 114
specific surface, 1–3, 7, 11, 13, 15, 22, 24–27, 29–30, 32–34, 36, 39–43, 47, 54, 56–58, 64–67, 69–72, 74–76, 82, 92, 96, 98, 128, 131–134, 142, 144–148, 150, 153–157, 166
specimen, 46, 85
steam, 50–52, 62, 84–85, 87
strength, 1, 3–4, 6–7, 14–17, 19, 22–24, 30, 33–40, 44, 46–58, 61–79, 82–106, 108, 114–125, 129–130, 134, 139–143, 145–153, 156–171
sulfate activator, 64, 69–70
sulfate-fluoride-alkaline activation (SFA), 66, 76
sulfate resistance, 4
superplasticizers, 5, 66, 134
surface activity, 5–6
surface energy, 5–6, 37–38, 164
surfactant, 5–6, 24, 38, 53, 155, 157, 164

T

technological factor, 12, 29, 33, 35, 37, 51–52, 55, 74–77, 82, 85, 93–94, 97, 99–100, 104–106, 115–117, 122–125
tensile strength, 4, 33, 46–48, 54, 99–100, 102, 104, 108
thermal expansion, 54
thixotropic, 15
tobermorite, 81, 130
tricalcium aluminate, 130, 132

U

unloading, 30

V

varied factors, 139
viability, 3, 13
viscosity, 2, 11–14, 30, 40, 55–56, 62, 134, 136–139
vitreous phase, 1–2, 7, 19, 62–63, 74
voidness, 3, 59

W

waste, 61, 63, 67, 96, 98
water demand, 2, 25–29, 33, 39–45, 57–58, 61, 67, 72–74, 79, 82–83, 85, 87, 89–90, 93, 95, 98–100, 113–116, 134, 143, 146, 148, 150–153, 165
  constancy rule, 40
water-cement ratio, 21, 40, 73, 97, 143
water-reducing effect (WRE), 82, 134
water-retaining, 3, 148
wettability, 6, 164
workability, 40–41, 57, 87, 89, 91, 129, 131, 134

X

X-ray diffraction analysis, 18, 79

Z

zeta potential, 5